Human Geography

Concepts and Applications

Ken Briggs

HODDER AND STOUGHTON
LONDON SYDNEY AUCKLAND TORONTO

British Library Cataloguing in Publication Data

Briggs, Ken
Human geography.
1. Anthropo-geography
I. Title
304.2 (expanded) GF41

ISBN 0 340 16116 7

First printed 1982
Fifth impression 1985

Printed in Great Britain for
Hodder and Stoughton Educational,
a division of Hodder and Stoughton Ltd.,
Mill Road, Dunton Green, Sevenoaks, Kent,
by Hazell Watson & Viney Limited,
Member of the BPCC Group,
Aylesbury, Bucks

Typeset by
Macmillan India Ltd., Bangalore

Preface

During the last 20 years great changes have taken place in human geography. The subject has evolved from a mainly descriptive approach towards a greater emphasis upon theory, and it is now concerned with the application and analysis of concepts and generalisations relating to patterns of location and spatial interaction. Expressions such as *locational analysis* have appeared and various spatial *models*, sometimes quite old, have been resurrected. Problems have been encountered as this new approach has filtered down into schools and examination syllabuses.

This book is intended to help to solve these problems. It approaches human geography from the 'modern' point of view, with sufficient breadth and depth to meet the demands of most modern GCE Advanced Level syllabuses. The exercises at the end of each major section are designed to provide the student with practice in essay writing and also in the answering of the *data-response questions* which some examining boards have adopted.

K. Briggs

Acknowledgements

The author and the publishers are grateful to the following, who supplied photographs for use in this book: British Leyland (p. 108); British Petroleum (p. 135); Camera Press Ltd (pp. 4 and 31); the Central Electricity Generating Board (p. 99); the Government of British Columbia (pp. 6, 79, 87, 90, 91, 95 and 171); the Information Service of Thailand (p. 24); the Japan Information Centre (p. 37); Robert Harding Picture Library (p. 36); Ross Hunter (p. 5); Michael Perris (pp. 160 and 170); Peterborough Development Corporation (cover); SNCF/Broncard (p. 134); and the United Nations (p. 30). The remaining photographs were taken by the author.

Contents

1 Population distribution and growth

1.1 World population distribution

DESCRIPTION OF THE PATTERN

The world's population is very unevenly distributed. In the first place, many more people live in the northern hemisphere than the southern. This is not surprising since the world's land areas are mainly in the northern hemisphere. Also, over three-quarters of the world's population live in Europe and Asia (the Old World). Relatively few live in the Americas, Africa and Australasia. Asia alone contains well over 2000 million people.

Figure 1.1 shows that there are four main concentrations of people. These are:

(*a*) Eastern Asia, which includes eastern China, Japan, Taiwan (Formosa) and Korea (containing over 1000 million);
(*b*) Southern Asia, which includes India, Pakistan, Bangladesh, Sri Lanka and Burma (a total of about 800 million in 1975);
(*c*) Europe, containing about 500 million excluding the USSR, but extending eastwards well into central Asia, and southwards to the Persian Gulf, the Nile Valley and North Africa;
(*d*) Eastern North America, comprising the eastern half of the United States and south-eastern Canada. This area contains rather fewer people than the other population clusters, having less than 200 million.

The two Asian population clusters are linked together by a densely populated coastal strip passing through Thailand, Kampuchea and Vietnam, and itself containing about 100 million people. The European cluster is separated from the Asian clusters by the relatively empty central part of Asia, and the North American cluster lies far away across the Atlantic Ocean.

The remainder of the world's population has a scattered distribution in offshore islands such as the West Indies, Java and the Philippines, and in small clusters around the coasts of the Americas, Africa and Australia. North of the Tropic of Cancer the major population clusters extend well into the centres of the two great land masses, but in the Tropics, and in the southern hemisphere generally, the pattern consists of a large number of relatively small clusters which rarely extend far inland.

THE CONCEPT OF POPULATION DENSITY

Figure 1.1 shows the parts of the world that have a population density of at least 10 people per square kilometre. Population density is calculated by dividing the number of people in a district by the area of the district in square kilometres. It shows the number of people who would be living in each square kilometre if people were distributed evenly over the land.

Within the population clusters shown in Figure 1.1, densities very much greater than 10 per

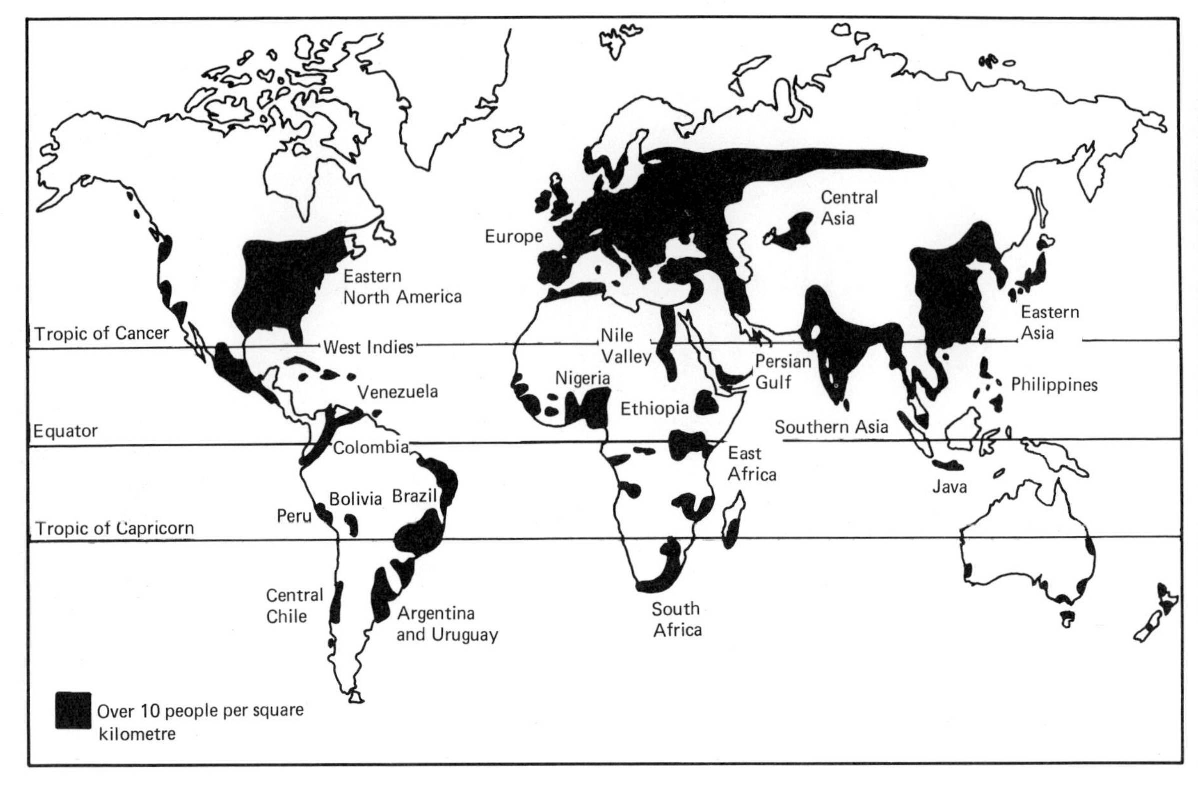

Fig. 1.1 World population distribution

square kilometre exist. Macao, for example, a Portuguese possession in Eastern Asia, has over a quarter of a million people living in an area of 16 sq km, giving a density of about 17 000 people per sq km. Other very densely populated small countries include Hong Kong, Gibraltar and Singapore.

Larger countries tend to have lower population densities for the country as a whole. Though they may contain small areas of very high density, large countries invariably contain relatively empty areas, thus reducing the average density for the country. The largest country in the world, the USSR, has a population of over 250 million, but its population density is only about 10 per sq km. Even China, with well over 800 million people, has a density of under 100 per sq km.

If very small countries like Macao are excluded, the three most densely populated countries in the world are Bangladesh (over 500 per sq km), and the Netherlands and Taiwan (both with over 400 per sq km). The United Kingdom has a density of a little over 200 per sq km.

FACTORS INFLUENCING WORLD POPULATION DISTRIBUTION

The distribution of population over the world appears to be related to both physical and economic factors.

Climate

Figure 1.2 illustrates some of the climatic factors that appear to have influenced world population distribution. In particular, few people live in areas where it is either excessively cold or excessively dry.

The northern coastlands of North America, Europe and Asia, as well as Antarctica, are cold for most of the year. Over large areas the mean temperature of the warmest month is under 10°C. Hence few crops can be grown and transport by both land and water is seriously hampered by frost and snow. The northern part of the Baltic Sea, Hudson Bay and much of the Arctic Ocean freeze in winter.

Crop cultivation is impossible in desert areas

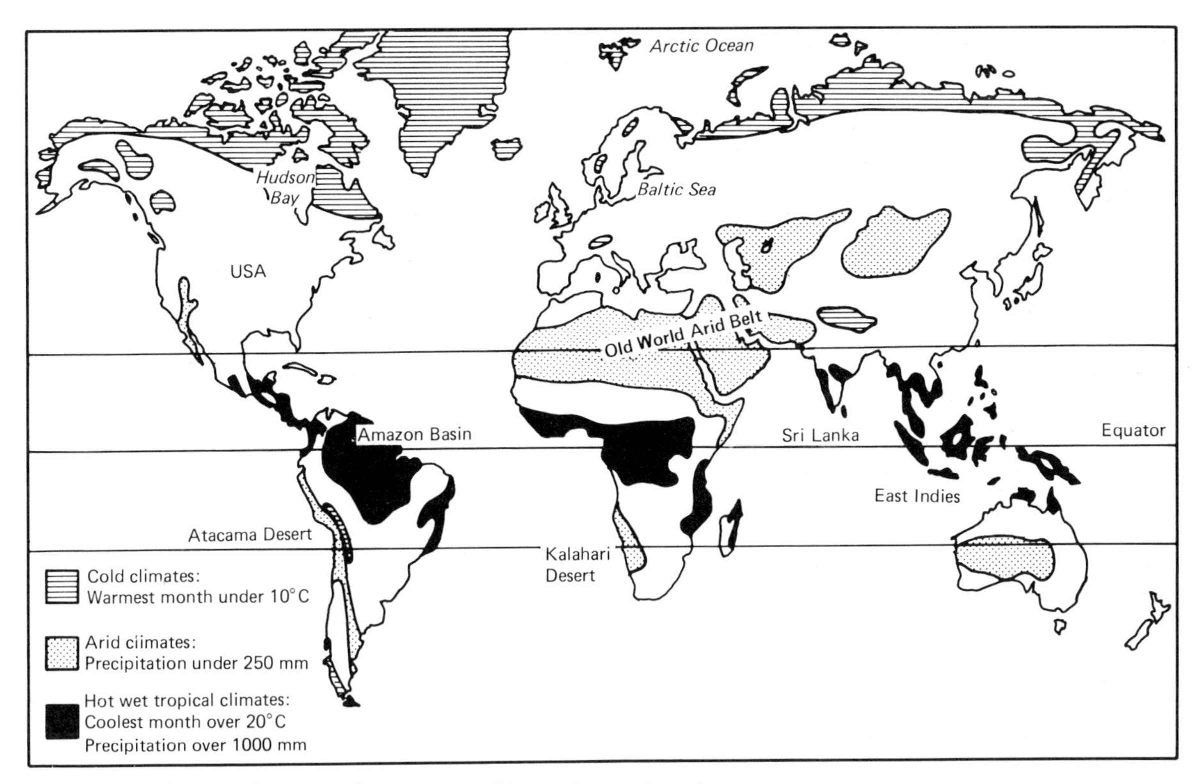

Fig. 1.2 Climatic factors influencing world population distribution

unless irrigation water is available. The great arid belt of the Old World extends almost continuously from north-west Africa, through Arabia, to north-west India and Mongolia, and divides the European population cluster from the two Asian clusters. Other arid lands include part of western USA, the Atacama Desert of Chile and Peru, the Kalahari Desert of south-west Africa, and the interior of Australia. Patches of relatively high population density do exist in arid areas where irrigation farming has been developed. Examples include the coastlands of Peru, the Nile Valley in Egypt and the Sudan, and parts of central Asia.

The climate of equatorial areas is often described as excessively hot and humid, and certainly large areas near the equator, such as the Amazon Basin and some East Indian islands, are very sparsely populated. On the other hand, some extremely hot, humid areas near the equator are very densely populated. Examples are Java (in Indonesia) and Sri Lanka. It appears therefore to be unsafe to conclude that man has tended to avoid living in the hot, wet, equatorial areas. Other quite different factors may have been involved.

A sparsely populated arid area in North Africa

The Himalayas – a sparsely populated upland area

Altitude

Figure 1.1 shows quite clearly that high, mountainous areas, such as the western cordilleras of North America, the Himalayas and the Plateau of Tibet, the Andes of South America, and the Alps and the Pyrenees in Europe, have low population densities. There are several possible reasons for this.

Man finds it difficult to breathe if he exerts himself at high altitudes. This is because the oxygen content of the air is very low. Special breathing equipment has had to be used in climbing very high mountains such as Everest. However, well below the level at which a shortage of oxygen becomes really serious, important climatic changes have taken place. Temperature decreases very rapidly with altitude. Hence high mountains, even on the equator, can be as cold as polar areas. Figure 1.2 shows some mountainous areas in which the mean monthly temperature is always under 10°C.

Mountainous areas also often have large areas of very steeply sloping land from which soil is easily eroded. Hence they are difficult to develop for agriculture.

In tropical areas, on the other hand, plateaux of moderate altitude often support relatively high population densities. Such plateaux are cooler and frequently considerably drier than the surrounding lowlands. Examples include the Andean plateaux of Venezuela, Colombia and Bolivia, and the plateaux of south-east Brazil, East Africa and Ethiopia.

Economic influences

Physical factors therefore are of some help in understanding the distribution of the sparsely populated parts of the world, but they are of little help with the densely populated areas. In fact the pattern of population densities over the world has a closer relationship with the types of economic activity shown in Figure 1.3. High densities of population tend to occur where productivity per unit area is relatively high. Four major types of economic activity may be considered:

(*a*) primitive subsistence economies;
(*b*) intensive subsistence economies;
(*c*) commercial ranching and grain cultivation;
(*d*) industry and intensive mixed farming.

In primitive subsistence economies people grow crops and rear or hunt animals mainly to support themselves, and there is little development of trade with other parts of the world. They mainly use only primitive techniques, and it takes a large area of land to support each person. Hence population density has to be very low. These economies are found mainly in three different parts of the world.

In the cold, high-latitude areas of the northern hemisphere, primitive groups such as the Eskimos and the Lapps live by hunting, fishing, trapping or the herding of animals. To some extent, advanced peoples have invaded these areas to exploit mineral and timber resources, as in the Canadian Shield, but mining and lumbering settlements form only very small patches of high population density in a vast, almost unpopulated area.

The great arid belt of the Old World is occupied by groups of nomadic animal herders who need to travel over great areas in order to provide pasture for their stock. Hence population density must be low. In certain areas, where substantial water supplies are available, irrigation works have been developed, resulting in high population densities in Egypt, Iraq and elsewhere.

In the hot, wet tropics (Fig. 1.2), in Africa, South America and the interior of south-east Asia, very primitive forms of crop production still survive. Shifting cultivation involves the culti-

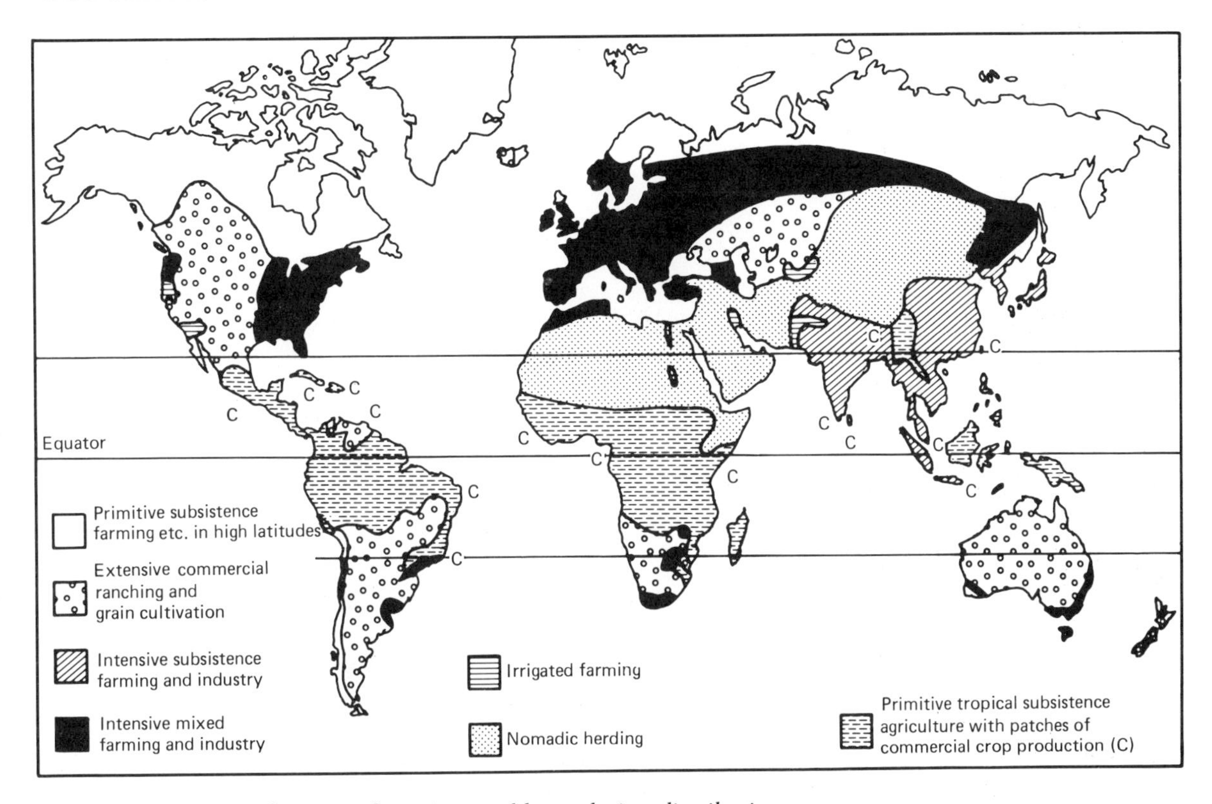

Fig. 1.3 Economic factors influencing world population distribution

vation of an area for several years until yields begin to diminish, followed by a movement to a new area of forest which is cleared by cutting and firing and cultivated for a few more years until another move is necessary. Such 'slash-and-burn' agriculture clearly can only support a low population density. In certain areas, however, commercial crop production has developed and land is permanently occupied. This gives higher population densities in the West Indies, eastern Brazil, Nigeria and East Africa for, example.

The rice farmers of Eastern and Southern Asia (Fig. 1.3) constitute intensive subsistence economies. Most production is for local consumption, and trade is relatively unimportant. These people have little labour-saving machinery, but they have developed farming techniques that produce a very high crop yield per hectare. Hence population densities are very high, particularly in the alluvial plains where irrigation systems have been established. Also many great industrial cities have developed, particularly in Japan.

Primitive farming in Thailand

A sparsely populated ranching area in Canada

Commercial ranching and grain cultivation dominate large areas of western North America, southern South America, central Asia, south Africa and Australia, usually in areas with relatively dry climates (Fig. 1.3). Development has been carried out mainly by people of European origin, and these areas are important suppliers of food to the industrialised regions of the northern hemisphere. Production involves the use of a great deal of capital equipment such as harvesting and cultivation machinery, all of which reduces the amount of labour that is needed. Hence population densities are low (Fig. 1.1). Output per hectare is generally quite low, but output per person (aided by machinery) is usually high. Hence living standards are generally high.

In Europe, parts of the USSR and eastern North America (Fig. 1.1), population densities are generally high because there exists a combination of industrial development in many cities and conurbations and highly productive farming involving both crop production and the rearing of animals. Often a great deal of both capital and labour are concentrated in small areas for the intensive production of dairy and market garden produce. In other areas fodder crops are grown to feed meat-producing animals. This type of economy has also been developed by European emigrants in the River Plate lowlands of South America, the Veld and Cape district in South Africa, and coastal areas of Australia and New Zealand. The large-scale development of intensive farming as described above is virtually restricted to areas which lie outside the tropics (Fig. 1.3).

1.2 Changes in world population

THE OVERALL GROWTH OF WORLD POPULATION

It is not known exactly when the first humans appeared but it was probably over $1\frac{1}{2}$ million years ago. For many thousands of years the number of human beings changed very little, and it had probably reached only about 250 million by the time the Christian era began. Population growth remained very slow for hundreds of years, and it is only in the last two hundred years that a marked increase in world population has taken place (Fig. 1.4). This has been associated with the occurrence of the Industrial Revolution in Europe and the subsequent extension of industrial and commercial activity to much of the remainder of the world.

The broad characteristics of world population growth are as follows:

(*a*) World population reached 1000 million in the early nineteenth century, 2000 million in the 1920s, 3000 million in the 1960s, and 4000 million in the mid-1970s. Thus each extra 1000 million has been added in about 100 years, 40 years and 10 years successively.

(*b*) World population doubled between about 1820 and 1920. It doubled again in about 50 years up to the 1970s. At the present growth rates it will double again in a little over 30 years.

(*c*) It is clear therefore that not only is the population of the world increasing; it is also increasing at an increasing rate.

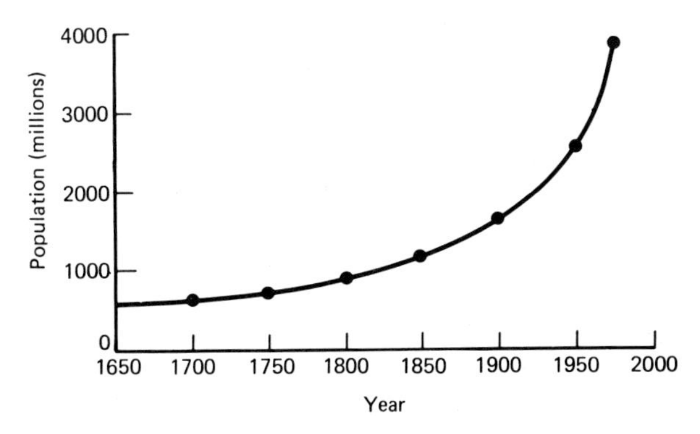

Fig. 1.4 World population growth

REGIONAL CONTRASTS IN THE RATE OF POPULATION GROWTH

This population increase has not taken place evenly over the whole world. Some areas have had much faster growth rates than others. Figure 1.5 illustrates some of these contrasts.

In the seventeenth century practically all of the world's five or six hundred million inhabitants lived in Asia, Europe and Africa (the Old World). Europe had about the same number of people as Africa.

Then the populations of Asia and Europe grew steadily, whereas Africa's population remained almost unchanged. By the early nineteenth century Europe's population total was double that of Africa. Throughout the whole period shown in Figure 1.5 Asia has had over half of the world's inhabitants.

In the nineteenth century population growth began to occur in North America and Latin America as they received immigrants from Europe and as industry and commercial farming began to develop. At first, population grew more quickly in North America, but in the twentieth century Latin

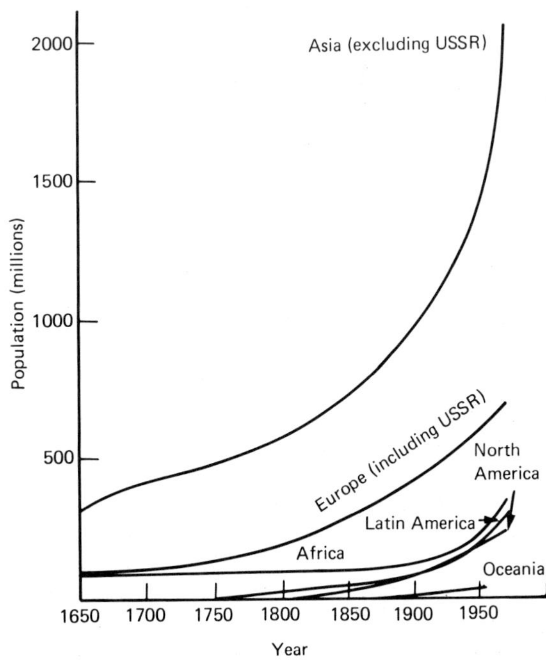

Fig. 1.5 World population growth by major regions

America has caught up and now has the greater total population.

In the twentieth century, especially since 1950, rates of population increase have been particularly high in Latin America and Africa (about 3% per annum). Asia's population has also grown fairly rapidly since 1950 but at a slower percentage rate than Latin America and Africa. However, sheer numbers of people in Asia are so great that even a moderate percentage increase means very large additions to the population total. As many as 400 million people have been added to Asia's population total in a period of 10 years.

In the twentieth century Europe has had the slowest rate of population growth of any continent.

The population of Oceania (Australia and the Pacific islands) is still barely 20 million, but its percentage rate of increase in the twentieth century has been higher than that of Europe. As in the Americas, this represents to some extent a migration of people from Europe.

Differences in the rate of population increase also exist within each continent. Figure 1.6 shows the countries that had the highest percentage rate of population increase in the early 1970s. It is clear that these countries whose populations have been growing very quickly are found mainly in Latin America and Africa. They are also countries which lie generally outside the main clusters of population shown in Figure 1.1. A large number of them are located in areas with either an arid or a hot, wet, tropical climate (Fig. 1.2). Also the great majority of these countries whose population is growing rapidly have relatively primitive economies such as nomadic herding or primitive subsistence agriculture (Fig. 1.3). It therefore appears that underdeveloped countries tend to be experiencing a faster rate of population growth than advanced industrial countries.

CAUSES OF CHANGES IN WORLD POPULATION

The population of an area changes according to the following principles. If the number of children

Fig. 1.6 Countries with a high rate of population increase

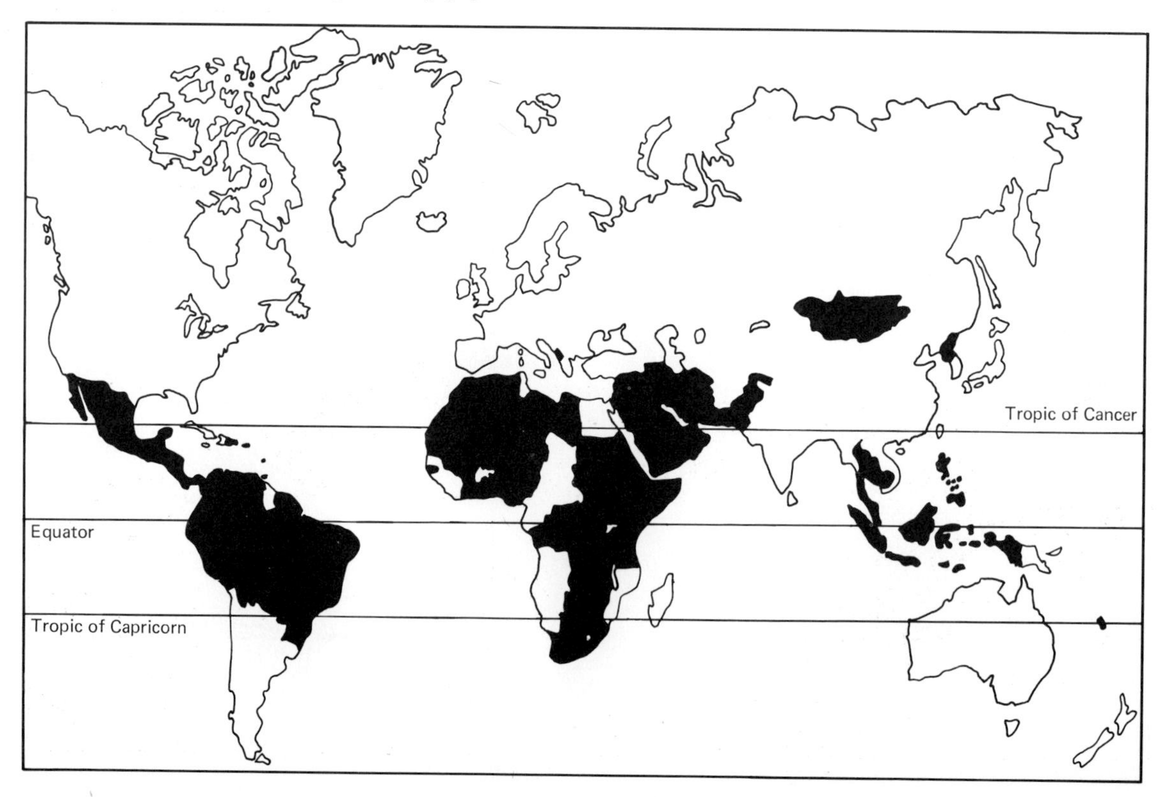

born in a particular year is greater than the number of people who die, the population tends to increase. This is called a natural increase of population. If the number of immigrants entering the area in a particular year is greater than the number of emigrants who leave, the population tends to increase.

The influence of the birth rate

The crude birth rate is the number of live births per 1000 people in a given year. It is calculated by dividing the number of births in the year by the number of thousands of people in the country. Variations in birth rates in different parts of the world are shown in Figure 1.7. If this map is compared with figure 1.6 it will be seen that the countries with the highest birth rates (Fig. 1.7) tend to be those in which population is increasing most quickly (Fig. 1.6). This correspondence is particularly close in the tropical parts of Latin America from Mexico to Brazil, in the Middle East, and in the islands off the coast of south-east Asia (including Indonesia and the Philippines). In Africa not all of the countries with high birth rates have rapidly increasing populations. This could be because their death rates are also high.

Birth rates are generally lowest in the advanced countries of Europe and North America.

The influence of the death rate

The crude death rate is the number of deaths per 1000 people in a given year. It is calculated in a similar way to the birth rate. Figure 1.8 shows the variations in death rates over the world. High death rates of over 20 per 1000 are almost restricted to Africa and parts of southern Asia. Death rates are generally low north of the Tropic of Cancer, throughout Latin America and in Australasia.

The rate of natural increase of population

The rate of natural increase of population is calculated by subtracting the death rate from the birth rate. The world distribution of the natural increase of population closely resembles the pattern shown in Figure 1.6.

Most of Africa has a natural increase of 20 to 30

Fig. 1.7 Variations in birth rates over the world

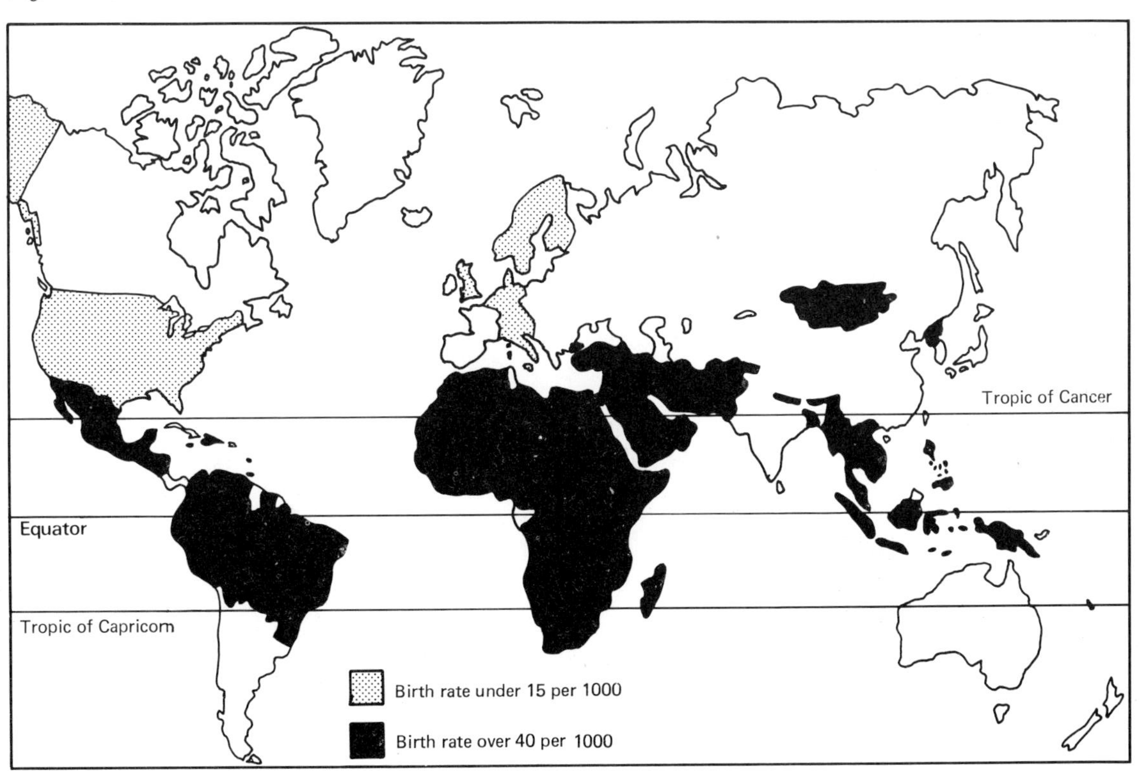

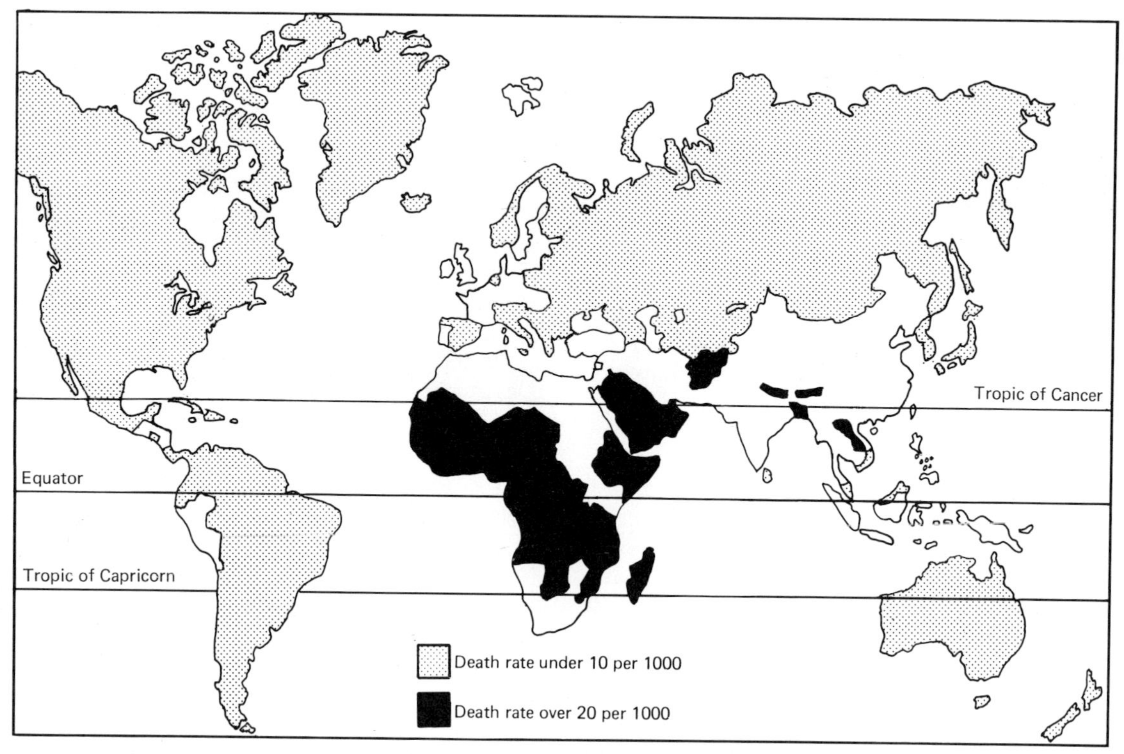

Fig. 1.8 Variations in death rates over the world

per thousand per year. Although death rates are high in some countries, birth rates are considerably higher.

In Latin America rates of natural increase are generally highest (over 30 per thousand per year) in central America from Mexico to Colombia, where a high birth rate (over 35 per 1000) is combined with a low death rate (under 10 per 1000). Natural increase is lowest in the West Indies and in Argentina, Chile and Uruguay, where the birth rate is much lower.

High rates of natural increase also exist in Thailand and the offshore islands of south-east Asia (over 30 per 1000).

Most of the remainder of the world has natural increase rates of about 10 per 1000 or under. These are mainly Europe (where sometimes the death rate is greater than the birth rate) and countries such as Canada, the USA, Australia and New Zealand which have been populated mainly by European emigrants.

The influence of population migration

The population map of the world owes a great deal to great migrations of people that have taken place in the past. The earliest parts of the world to become densely populated were India, China and Europe. From these areas people have migrated into the less densely populated areas. Chinese or Indians for example have moved into south-east Asia, eastern and southern Africa and the Americas (Fig. 1.9). The greatest migration of all has been the movement of Europeans into the relatively empty lands of the Americas, South Africa and Australasia. Well over 100 million emigrants have left Europe for these destinations.

Before the nineteenth century European emigration was on a small scale and consisted mainly of the setting up of farming colonies, trading stations and plantations on the coasts of the Americas, Africa and southern Asia. In the nineteenth century however steamships made ocean travel much faster and a rapid growth of population took place in Europe (Fig. 1.5). At this time Europeans occupied many sparsely populated areas of the world.

The greatest flow has been to the USA, totalling about 40 million emigrants. For most of the

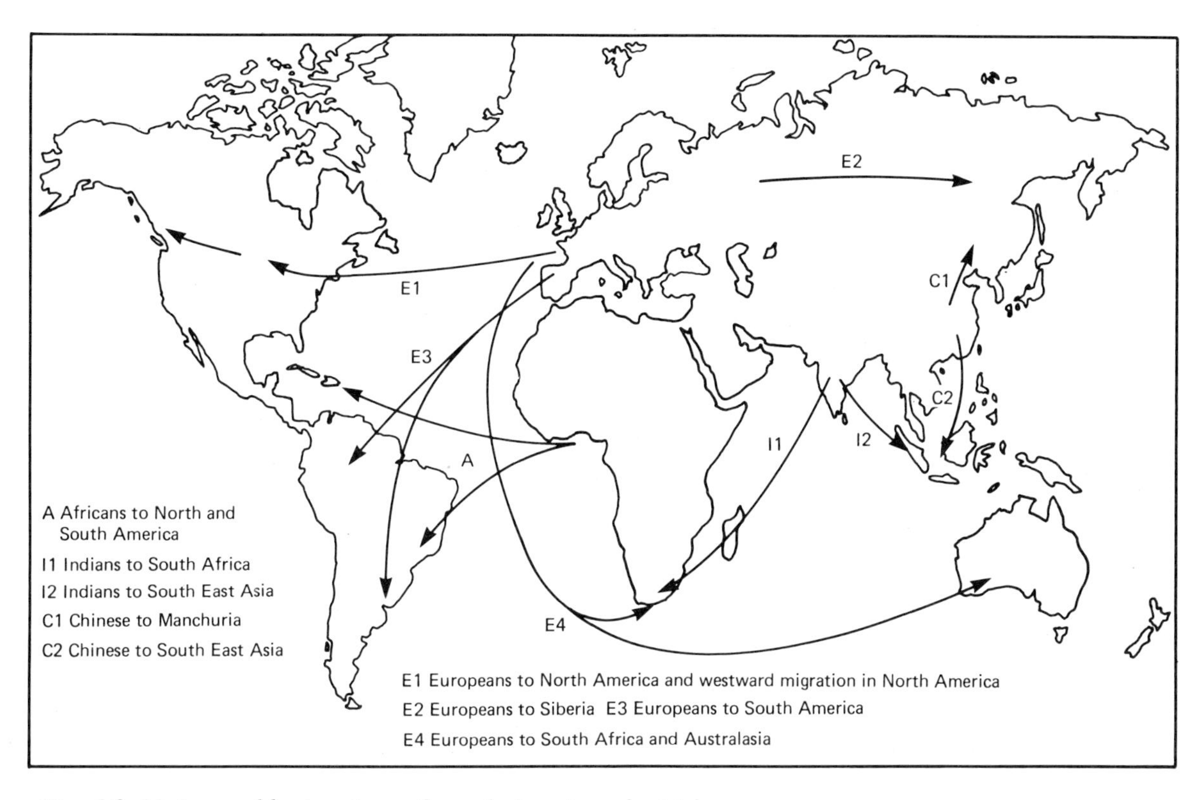

Fig. 1.9 Major world migrations of population since the 16th century

nineteenth century these emigrants came mainly from Britain and Germany whose populations were growing as they became industrialized, but large numbers also emigrated from Ireland and from Scandinavia. Later in the century greater numbers of emigrants began to come from southern and eastern Europe, particularly Italians, Greeks, Hungarians and Russians. Italy alone lost nearly 20 million emigrants in the 40 years before the First World War. After the First World War, the USA began to restrict immigration and this migration from Europe decreased considerably. In more recent years, the USA has received many migrants from Latin America, especially from Mexico and Puerto Rico. Another large-scale migration in which the USA was involved was the slave trade, a forced migration of Africans to the plantations of southern USA (Fig. 1.9).

South America, too, was largely populated by European emigrants. These were mostly Spaniards, Portuguese and Italians. Argentina, for example, received about six million European immigrants in about 50 years round about the end of the nineteenth century, and Brazil has received many Italians and Portuguese, in addition to African slaves.

After the First World War emigration from Europe decreased considerably. Some countries introduced restrictions on immigration and, in any case, population growth in Europe was slowing down. Nevertheless, about a million European Jews emigrated to Israel after the Second World War, and several million Europeans have emigrated to Australasia since 1945.

1.3 The demographic transition

Historical records show that population totals in some European countries have changed in the last two centuries according to an easily recognisable pattern. For most of the eighteenth century, despite a very high birth rate, population numbers were kept low by the very high death rate. During the nineteenth century, a striking decrease in the death rate took place, while the birth rate generally remained high. Thus natural increase produced a rapid growth of population. In the twentieth century, death rates have remained low, but the birth rate has tended to decrease. Thus the rate of population growth has gradually decreased until, at the present time, the population total is relatively stable at a high level. It has been supposed that the population totals of other countries, and their birth and death rates, will change in a similar way as they develop economically. The model which describes this sequence of population changes is known as the demographic transition (Fig. 1.10).

THE FOUR STAGES OF THE MODEL

Stage 1

In a primitive, subsistence economy, the birth rate and the death rate are both high, so there is little tendency for the population to increase. The birth rate remains fairly stable from year to year, but the death rate may change erratically from time to time, often as a result of famine, war or disease. Wars lasting only a few years have frequently cost millions of lives. Disastrous famines have taken place even in modern times. The population total therefore fluctuates considerably, but remains generally at a low level.

Stage 2

In stage 2 a steady decrease in the death rate occurs and life expectancy increases, but the birth rate remains at its previous high level. Hence population increases at an increasing rate. Famines become less frequent as food supplies increase and become more reliable. Greater political stability reduces the occurrence of wars and, perhaps above all, improved medical facilities and sanitation reduce the incidence of disease.

Stage 3

In stage 3, the birth rate in turn begins to fall, as the society becomes urbanised and industrialised. Hence the population total continues to increase but only at a decreasing rate. By this stage, infant mortality has decreased to a low level, so there appears less need to produce children. Birth control techniques become available. Also educated middle classes may have evolved who are concerned to improve their living standards and may regard the rearing of children as a hindrance to these ambitions. Women are no longer tied to the home, but begin to take up professions, in which they may feel that children would restrict their progress.

Stage 4

By stage 4 birth rates have fallen to approximately the same level as death rates and population growth has more or less ceased, but the population total is now higher than it has ever been. At this stage, in contrast to stage 1, the birth rate tends to fluctuate rather more than the death rate.

It is by no means certain that underdeveloped countries will pass through the ideal stages of this

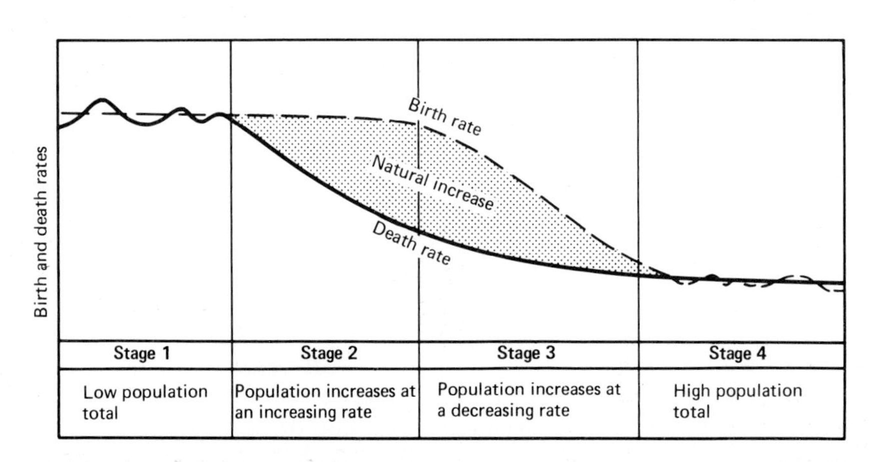

Fig. 1.10 The demographic transition

model. There seems little doubt that death rates will steadily decrease, as in stage 2 of the model, because the facilities of modern medicine already exist and only need to be passed on to developing countries. They tend to be accepted readily, because most people wish to delay death as long as possible. Death rates have been reduced most successfully in small, densely populated islands, where the population is easily accessible, such as Singapore, Barbados and Puerto Rico. The death rate in Puerto Rico, for example, was halved in well under 40 years. A reduction in the birth rate seems to take place much less readily. It may be difficult to persuade the population that fewer births would be beneficial, and problems have been met in extending education in the use of birth control techniques.

THE WORLD PATTERN OF DEMOGRAPHIC STAGES

In the light of current birth and death rates it is possible to map the world distribution of the four demographic stages, though it is by no means certain that future developments will conform exactly to the model.

Stage 1

Countries with high birth and death rates are concentrated in Africa and southern Asia, particularly in comparatively isolated areas such as Afghanistan, Nepal, Bhutan and Laos, and in areas where European settlement has been rare, such as the lower lands of central Africa in Zaire, Nigeria and Ghana (Fig. 1.11). Even in this group of countries birth rates are generally higher than death rates, but their death rates are particularly high in comparison with those of the rest of the world, as Figure 1.8 shows. According to the demographic transition model we would expect these countries soon to experience a population explosion. In fact rapid population expansion is already a reality in some of them (Nigeria, Ghana, Ethiopia, Tanzania and Saudi Arabia, for example) but in others, such as Angola, Afghanistan and Bangladesh, population growth is slower (Fig. 1.6).

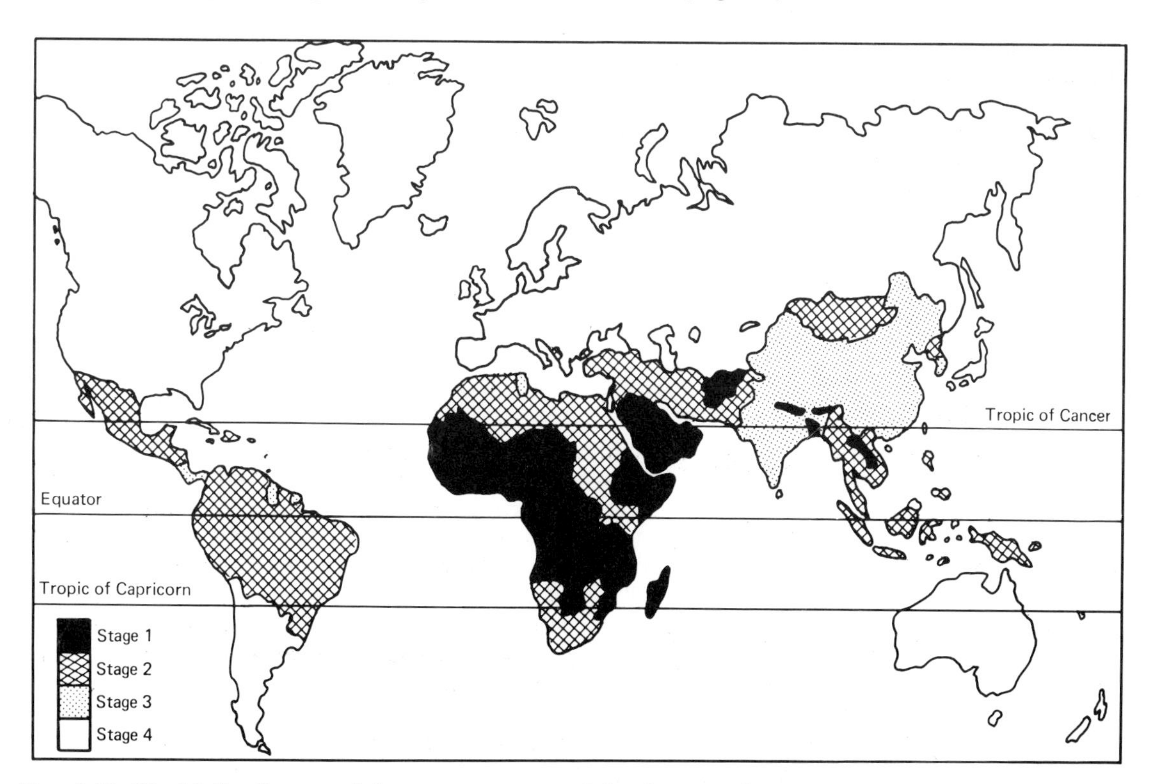

Fig. 1.11 World distribution of the various stages of the demographic transition

Stage 2
In the group of countries at stage 2 birth rates remain high but death rates are distinctly lower (under 20 per thousand). Population growth rates are high almost everywhere (Fig. 1.6) and are particularly so in the American tropics from Mexico to southern Brazil. Populations are also growing fast in parts of Africa, the Middle East, Pakistan and the East Indian islands.

Stage 3
In stage 3 the birth rate has begun to fall and in this group of countries it stands at 25–35 per thousand. According to the model, population growth should still be taking place but the rate of growth should be slowing down. Some examples are found in the American tropics, such as Costa Rica, Panama, Jamaica, Surinam and French Guiana, but China and India are the most populous members of this group.

Stage 4
In the countries at stage 4 the population is frequently increasing, but only at a slow rate (Fig. 1.6). These countries comprise most of Europe, the USSR and Japan, together with other temperate lands in both hemispheres which have been occupied and developed mostly by European emigrants. Examples are Canada and the USA, Chile, Argentina, Uruguay and Australasia. A few West Indian islands, such as Cuba, Barbados, Puerto Rico and Trinidad are included in this group. Their birth rates are slightly higher than the European average, but their death rates are remarkably low.

THE EFFECT UPON POPULATION PYRAMIDS

A population pyramid is a diagram that describes the age and sex distribution of a country's population. It consists of a vertical axis from which extends a set of horizontal bars. The length of each of these bars shows the percentage of the country's population that is contained within a particular age range, which may be five years or ten years, as in this chapter. Male population is represented to the left of the vertical axis and female population to the right. In this chapter we are considering only the effects of the birth and death rates on the shape of the pyramid. Migration will be considered later.

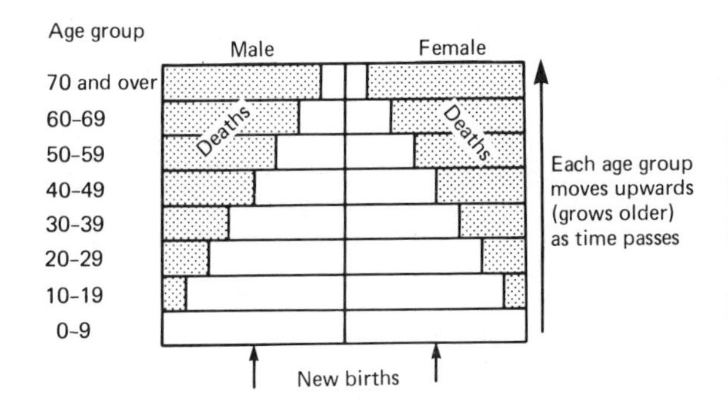

Fig. 1.12 A population pyramid

A population pyramid is bound to be approximately triangular in shape, thinning upwards towards the older age groups of the population. This is because, as time passes, the numbers of people in each age group will inevitably decrease as some of them die. New births of course can only enter the pyramid at the lowest age group. This process is illustrated in Figure 1.12.

The effect of a fall in the death rate
Figure 1.13(*a*) shows an imaginary pyramid in which only half of each age group survives each 10-year period. Thus, each successive age group has only half the number of members as the next younger group. Very few people survive to the age of 70. The 0–9 age group contains half of the total population.

In Figure 1.13(*b*) the same number of people exist in the youngest age group, but the death rate has now fallen, and only a quarter of each age group (instead of a half) dies in each 10-year period. Clearly the total population has increased, and the pyramid is very much broader than before. The youngest age group now contains much fewer than half of the total population. It should be remembered though that it would take at least 70 years for the complete pyramid to obtain its new shape. The narrow 'peak' in the older age groups would last for many years, despite the changes taking place in the younger groups. A population pyramid is in fact a record of the country's population history ever since the birth of the oldest inhabitant.

The effect of a fall in the birth rate
Suppose now that the death rate remains as in Figure 1.13(*b*), but that fewer births occur, so that

the 0–9 age group becomes smaller. This could produce a narrowing of the youngest age group, as shown in Figure 1.13(*c*), but older age groups will not be affected until this relatively small age group makes its way up the pyramid. Figure 1.13(*d*) shows that it may be many years before this reduction in births begins to affect the upper end of the age range.

Examples of population pyramids

The above explanation is very much simplified, but it may aid the understanding of the significance of the shapes of the population pyramids of a variety of countries. One slight complication is that population pyramids usually show the percentage of the total population in each age group and not the actual number.

Figure 1.14 shows population pyramids for Bangladesh, Ecuador, Argentina, Japan and France in the mid-1970s. They give an excellent impression of the transition from stage 1 to stage 4 of the demographic transition.

Bangladesh has been classified as of stage 1 in Figure 1.11. It has a birth rate of well over 40 per thousand and a death rate approaching 30 per thousand. Its pyramid clearly has the thinnest peak in the over 60 age group. Well under 5 % of the population of Bangladesh have reached the age of 60, whereas well over 50 % are under the age of 20.

The pyramid for Ecuador is rather similar to that for Bangladesh. Ecuador still has a birth rate almost as high as that of Bangladesh, but its death rate has recently fallen to a very low level. Well over 50 % of its population are under the age of 20, but a greater proportion has reached the age of 60 than in Bangladesh. Ecuador has therefore been classified as a member of stage 2 of the demographic transition (Fig. 1.11).

Argentina's pyramid is quite different. Its base is much narrower and its peak is much broader. Its birth rate is less than half that of Bangladesh, but it is still over 20 per thousand. Hence, Argentina appears to have moved from stage 3 into stage 4 very recently. It has a death rate of about 10 per thousand, and over 10 % of its population are aged 60 or over.

Japan's birth rate and death rate are both lower than Argentina's so it seems reasonable to classify it as of stage 4. As recently as the 1940s Japan's birth rate was well over 30 per thousand, but since 1950 the introduction of sterilisation and the

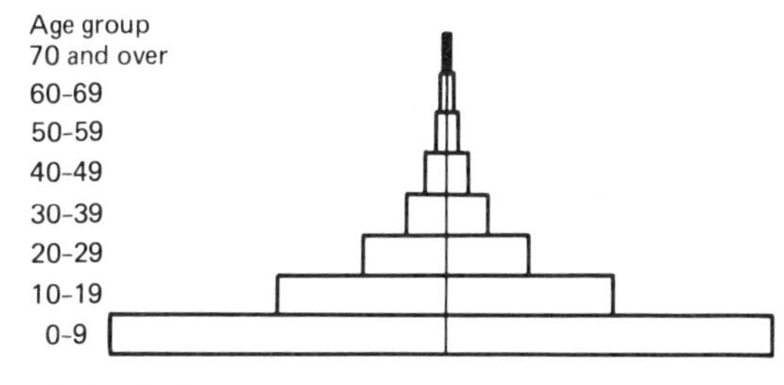

(a) High death rate

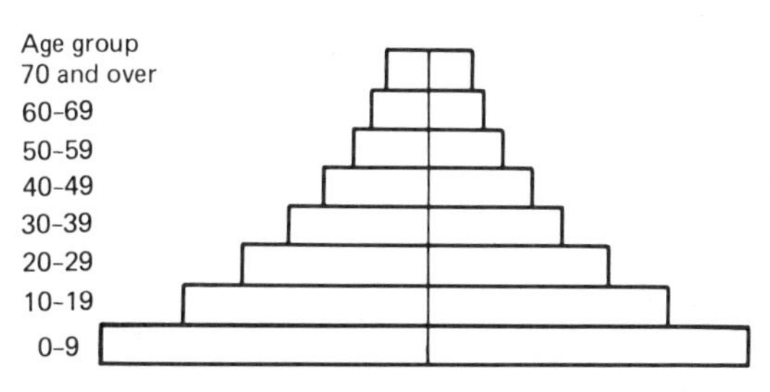

(b) Reduction in death rate

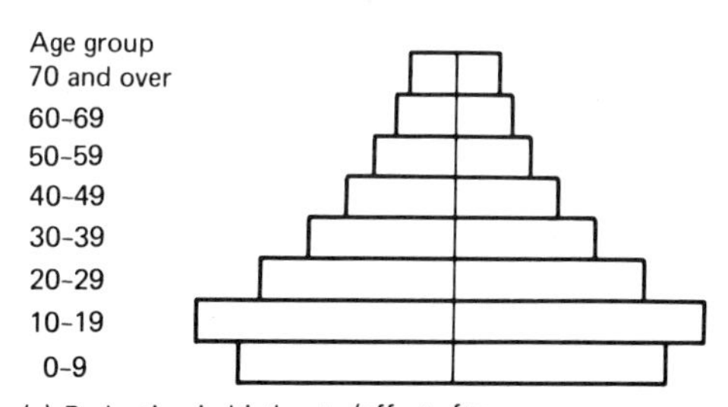

(c) Reduction in birth rate (effect after 10 years)

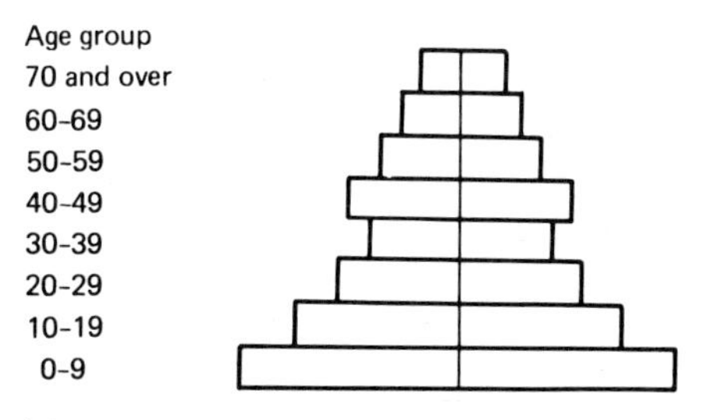

(d) Reduction in birth rate (effect after 40 years)

Fig. 1.13 Theoretical examples of population pyramids

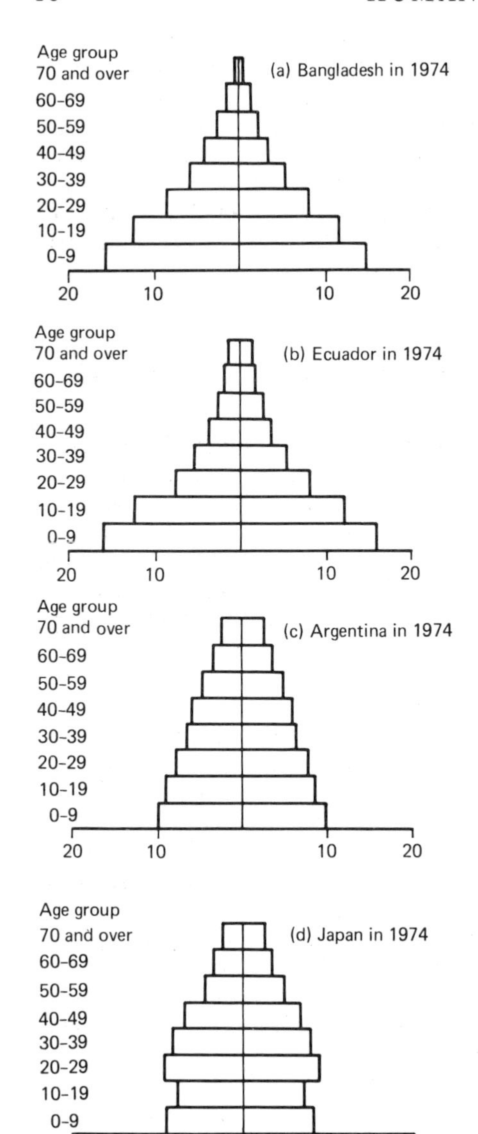

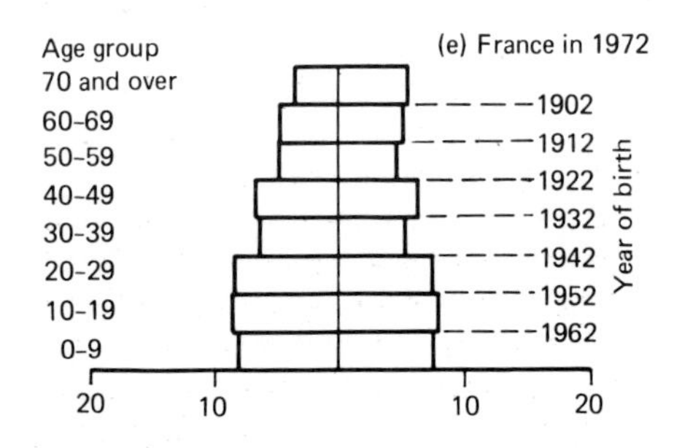

Fig. 1.14 Actual examples of population pyramids

spread of contraception have reduced it to under 20 per thousand. Thus Japan's pyramid becomes narrower at the 10–19 age group. The members of this group were born in the late 1950s and the early 1960s. Japan's rate of natural increase of population is now quite low. Well over 10 % of Japan's population have reached the age of 60, and only about 30 % are aged under 20. Of the five countries illustrated in Figure 1.14, Japan has the largest proportion of people in the 20–60, economically active, age range.

France appears to have advanced further into stage 4 than Japan. Its birth rate is lower than Japan's, and its population has virtually ceased to grow. Very nearly 20 % of France's population have reached the age of 60 and these include a particularly large number of females. However, France's population pyramid shows other interesting characteristics. In three cases a younger age group contains fewer people than the one that is immediately older. One of these is the group made up of people aged 50 to 59, who were born between 1912 and 1922, a period in which the First World War occurred, when France suffered heavy casualties and the birth rate declined. The next relatively small age group is that born between 1932 and 1942, whose members are aged 30 to 39. This period of course contains the Second World War, and it is also the time at which the relatively few people born between 1912 and 1922 have reached normal childbearing age. Finally, the youngest age group of all, born between 1962 and 1972, is also relatively small. This is the approximate time at which people born round about the time of the Second World War have reached childbearing age. Thus, although France's declining birth rate matches the principles of the demographic transition, it is clear that the influence of two wars has been of great importance.

1.4 The mobility transition

As the changes in population take place through the demographic transition it appears that changes also take place in the tendency of people to migrate either by going to live in a different place or by making journeys to visit other places temporarily. Four stages of this so-called 'mobility transition' have been identified, and these appear to correspond approximately to the four stages of the demographic transition.

THE FOUR STAGES OF THE MOBILITY TRANSITION MODEL

Stage 1

In a primitive, subsistence economy, people move about very little. They make daily journeys to work in the fields or to fish. Occasionally rather longer journeys may be made, to visit a shrine, for example. But the majority of people have their homes in the district in which they were born.

Stage 2

In stage 2 of the demographic transition the death rate begins to fall and population begins to increase at a fast rate. At this stage it appears that a strong tendency to migrate appears. People begin to move their homes either to other parts of their own country or to other countries. Some may move into relatively empty parts of their own country and set up farming settlements there; others may move to live in cities not far away. Many emigrate to more advanced countries which they believe offer better opportunities. At this stage migration to other countries and to cities reaches its maximum importance.

Stage 3

As the birth rate falls and population growth slows down, overseas emigration tends to decline, but movement from the countryside to the cities tends to remain important. Also, a good deal of internal migration from one city to another and within particular cities tends to take place.

Stage 4

In an advanced society in which the birth rate has fallen so that the population total remains almost constant, migration from the countryside to the cities is much reduced, but people tend to move home frequently either from one city to another or from one district of a single city to another. About one-fifth of the population of the United States change their home address annually. Travel is now so easy that many long, temporary journeys are made. Many travel long distances every day to work. Young people move in order to attend a college or a university, or to find a job, or to get married. Journeys of thousands of miles are made by air in order to enjoy holidays. At this stage, people are more mobile than ever before, but they rarely emigrate to live in another country.

Some think that this stage will be followed by a further stage in which people change their homes less often, but make many long, temporary journeys. The development of Concorde, for example, makes it possible to make a day-trip from London to New York.

THE WORLD PATTERN OF MOBILITY STAGES

Much of Europe has passed through all four stages of the mobility transition since medieval times. As the death rate declined in Europe in the nineteenth century (stage 2 of the demographic transition), great international migrations of population took place (stage 2 of the mobility transition). In the early part of the nineteenth century migrants to the United States came mostly from north-west Europe, the first area to pass into stage 2 of the mobility transition. Later, more migrants began to come from southern and eastern European countries as they entered stage 2. In the twentieth century, large numbers of immigrants into the United States have come from Latin America. It appears therefore that stage 2 of the mobility transition has spread outwards from north-west Europe to those areas that were influenced by Europeans or European ideas. The later stages of the mobility transition have been reached in those areas where European influence has been strongest, such as the United States, Australia and Japan.

Taking the world as a whole, relatively immobile (stage 1) societies are now virtually restricted to Africa. Examples of countries that have passed into stage 2 are most Latin American countries from Mexico to Brazil, from which many migrants travel to live in the United States, and within which cities, such as Mexico City and Sao Paulo, are growing at an enormous rate. India, Pakistan and Bangladesh are important sources of migrants to the United Kingdom, and here, too, cities are expanding at a rapid rate. Most of Europe, the USSR, North America, southern South America and Australia and New Zealand may be regarded as having reached stage 4 of the mobility transition, just as they have reached stage 4 of the demographic transition (Fig. 1.11).

THE EFFECT OF MIGRATION UPON POPULATION PYRAMIDS

People who migrate from one country to live in a different country are predominantly young adults, and are mostly single men or childless married couples. Hence migration tends to have two main effects upon the shape of population pyramids.

The sex distribution

Countries that have received many migrants tend to have a greater number of males than females in their population. This applies particularly in remote parts of countries such as Canada and Australia.

The age distribution

Large numbers of immigrants tend to widen the base of a country's population pyramid, especially below the age of 30. In British Columbia, for example, between 1961 and 1971, the population total increased by about a third, mostly by immigration, since the birth rate was declining. During this same period the number of people aged under 30 increased by over 40 %. Such a great increase in the proportion of younger people in the population reduces the proportion of people in the older age groups, thus tending to make the pyramid narrower at the top. This could give a false impression that the death rate was increasing.

Similarly, a country that has lost large numbers of young people through emigration will tend to have a pyramid with a narrower base and a wider peak. The increase in the percentage of older people is purely the result of the decrease in the *number* of younger people.

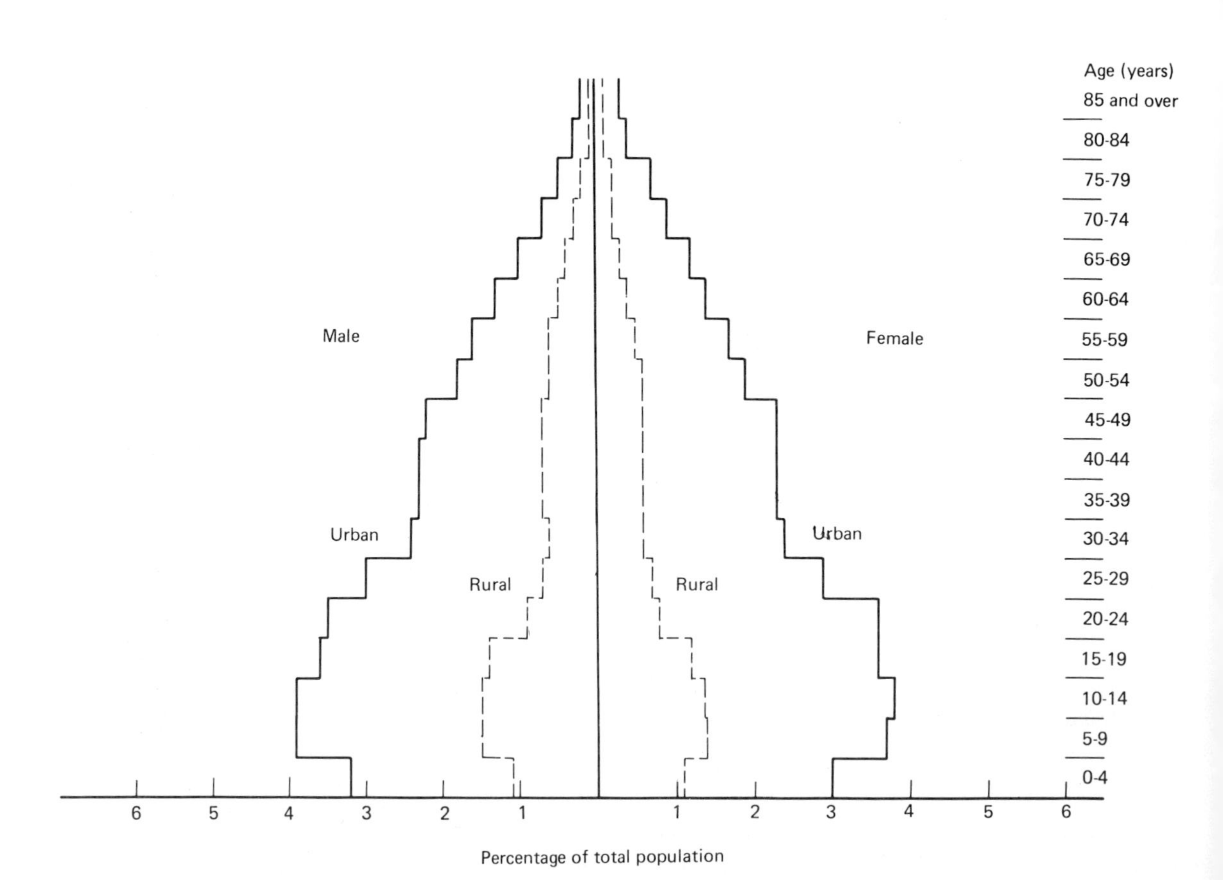

Fig. 1.15 Age-sex distributions of Canada's urban and rural populations (1971)

Exercises

1. Discuss the truth of each of the following statements, quoting examples wherever they are appropriate.

(*a*) Relief and climate are the major influences on the distribution of population over the world.
(*b*) Population is increasing most rapidly in those parts of the world that already have a high density of population.
(*c*) As the mobility of population increases international migration tends to decrease.

2. (*a*) In which countries at the present time is there a very strong tendency for population to migrate overseas?
(*b*) According to the mobility transition model, which countries would you expect to provide most overseas emigrants in the remainder of the twentieth century? Justify your choice.

3. Discuss the extent to which advanced industrialized countries have higher national population densities than developing agricultural countries. Suggest reasons for the differences that you identify.

4. Figures 1.15–1.17 show population age–sex pyramids for the urban and rural populations of Canada, Norway and Zambia. On each Figure the urban and rural populations are shown separately. The total population is the sum of these two elements.

(*a*) Which of these countries has the greatest proportion of its inhabitants aged (i) at least 70? (ii) under 10? (iii) 20–49 inclusive?
(*b*) For each country describe the differences between the age–sex distributions of its urban and rural populations, and explain the reasons for these differences.
(*c*) Which stage of the demographic transition has been reached by each of these three countries? Justify your choice.

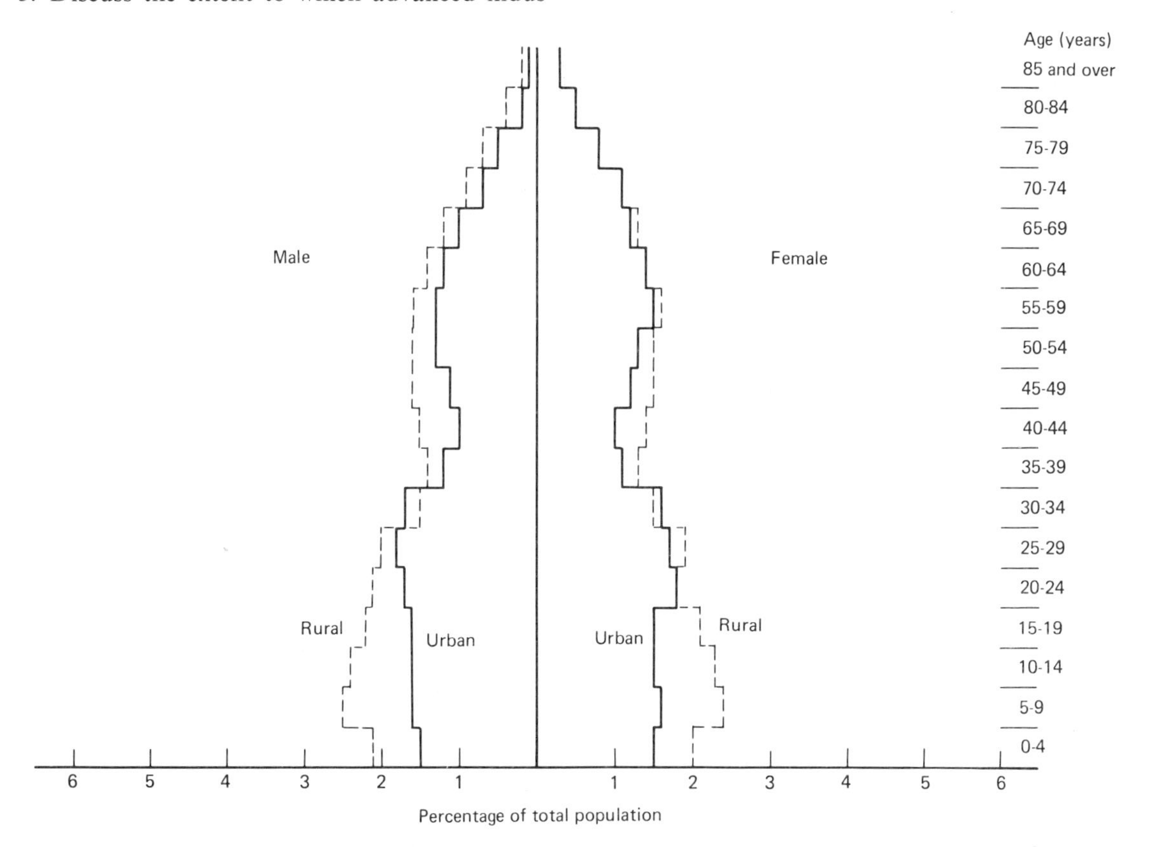

Fig. 1.16 Age-sex distributions of Norway's urban and rural populations (1977)

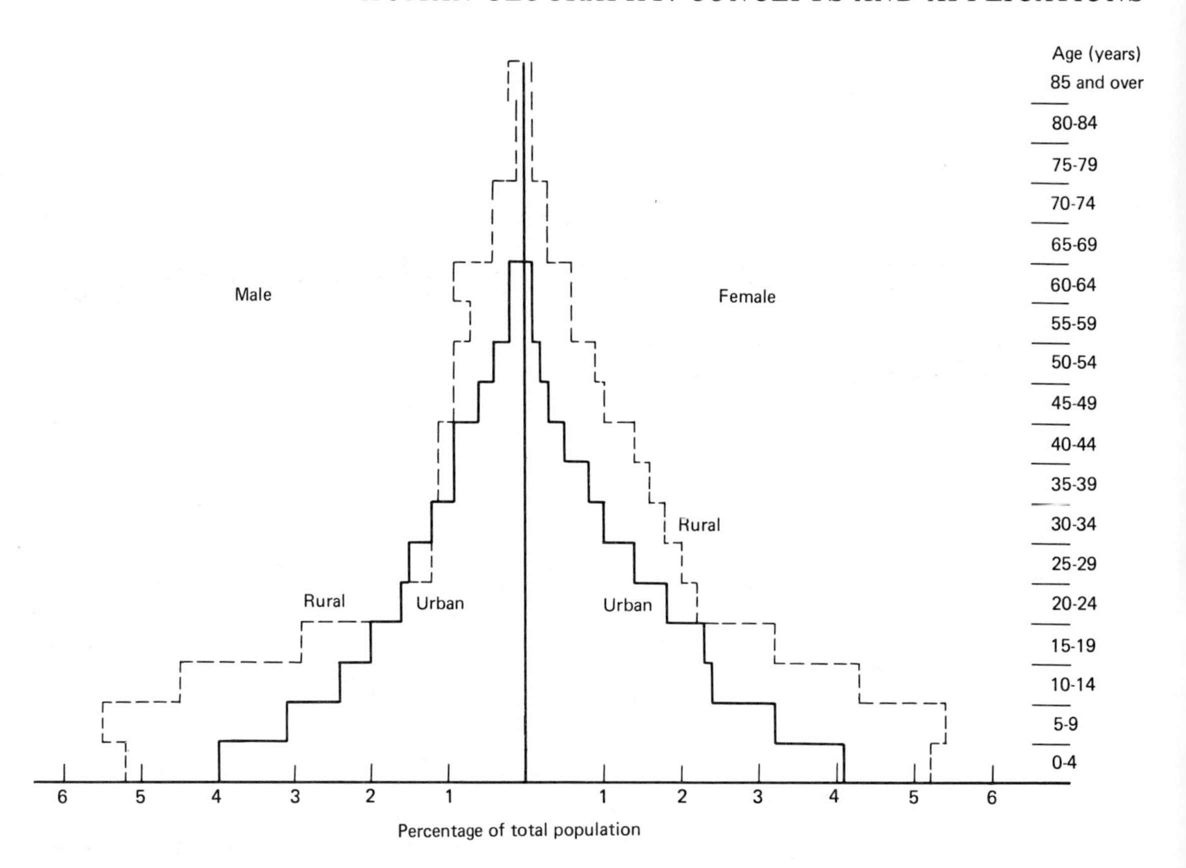

Fig. 1.17 Age-sex distributions of Zambia's urban and rural populations (1974)

(Data derived from UN Demographic Yearbook, Special Issue, 1979, pp. 244–5, 258–9. Copyright, United Nations (1979). Reproduced with permission.)

2 Characteristics of the world's population

2.1 Human races

All human beings are members of a single species known as *homo sapiens*, and are descended from the same source. It is obvious however that great differences exist in the appearance of people who live in different parts of the world. The colour of their skins for example may be white, pink, brown, black, yellow or bronze, together with many intermediate shades. Also peoples' hair may be straight, wavy or woolly with many detailed variations. It is also known that blood types vary between different groups of people. Nevertheless, all the different kinds of human beings are capable of interbreeding and producing children. These different groups of human beings are known as 'races' or 'ethnic types'.

THREE MAJOR RACIAL (ETHNIC) TYPES

It is really impossible to divide mankind into distinct racial types since every individual person is different, but usually three broad groups of human beings are recognized:

(*a*) Caucasoid peoples have fair skins and wavy hair.
(*b*) Mongoloid peoples have yellow or brown skins, straight black hair and high cheek-bones.
(*c*) Negroid peoples have black or dark brown skins, black woolly hair and thick lips.

However, these are very inadequate descriptions. Most people do not match any single one of these descriptions. The Caucasoid group, for example, includes tall, fair haired Scandinavians as well as shorter, darker skinned Indians.

PROBABLE ORIGINS AND EARLY MIGRATIONS

Man appears to have originated in the Old World at some time during the Ice Age. At this time movement would not be easy, and groups of people could live in isolation for long periods of time, so that many generations of interbreeding took place. This is believed to have produced the three major racial types listed above. Differences in skin colour, for example, could have been the result of adaptation to climate, possibly the intensity of sunlight. As time went by, great drifts of population seem to have taken place.

The Caucasoid peoples appear to have moved from their homeland in south-west Asia, near the Caspian Sea, north-westwards into Europe, south-westwards into north Africa, and south-eastwards into India. Westward movement was of course limited by the Atlantic Ocean.

The Mongoloid peoples seem to have moved from central Asia, where they evolved, southwards to India, south-eastwards to China and south-east Asia, and north-eastwards to enter North America via the Bering Strait. They then appear to have spread southwards into South America.

Negroid peoples probably originated in north-central Africa, and then migrated into southern Africa.

Movements from one land area to another would be helped by the fact that the level of the sea was lower at times during the Ice Age than it is at the present day.

Thus, Europe, the Middle East and India were occupied by Caucasoid peoples; most of Asia and the Americas by Mongoloid peoples; and central and southern Africa by Negroid peoples.

LATER MIGRATIONS AND THE PRESENT DISTRIBUTION

Since the 'Great Age of Discovery' about 500 years ago, when the Caucasoid peoples of Europe 'discovered' the Americas, southern Africa and Australasia, large-scale migrations across the world's oceans have become possible. Thus European Caucasoid peoples have migrated in large numbers to the Americas, southern Africa and Australasia. These migrations are described on pages 10 and 11.

The result of these large-scale migrations was that 'multi-racial' or 'plural' societies came into existence. Caucasoid peoples met Mongoloid peoples (American Indians) in North and South America. Intermarriage produced Mestizos in South America. Caucasoid Europeans met Negroid peoples in southern Africa and, here again, interbreeding took place.

The slave trade, organized mainly by Europeans, transferred large numbers of Negroid Africans to North and South America. These movements are shown in Figure 1.9.

PROBLEMS OF MULTI-RACIAL (PLURAL) SOCIETIES

Apartheid in South Africa

The population of South Africa consists mainly of Africans ('blacks') who make up 70 % of the total, but political and economic power is in the hands of a relatively small number (17 % of the total) of Europeans ('whites'), mainly of British, Dutch, French and German descent. In addition there are smaller numbers of 'coloureds' (9 %) and Asians, mainly Indians (3 %).

The Africans were originally subsistence farmers but, as South Africa developed economically, they tended to migrate to the main cities where they now mostly work at unskilled jobs for low wages. Soweto, near Johannesburg, is an African township with well over half a million inhabitants. The coloureds also live mainly in urban areas, especially in Cape Province. Indians came originally as indentured labourers to work on the sugar plantations of Natal.

Soon after the Second World War the South African government adopted the policy of apartheid—a policy of separate development for each of South Africa's 'nations'. Social contact was to be reduced especially between 'whites' and 'non-whites' by providing separate residential areas and other means. Mixed marriages were made illegal. Identity cards were issued stating the owner's 'race'. Also 'homelands' or 'Bantustans' were created in which Africans could develop 'separate identities'. These areas roughly corresponded to the areas occupied by the various African tribes before they began to migrate to the cities. The two largest are the Transkei and Kwa Zulu, to the south and north of Durban, respectively. In the long run, Europeans, coloureds and Asians would move out of these homelands, and Africans would take their places.

Many problems exist in the creation of the homelands. Farming is difficult because of the hilly relief of the eastern areas and the low, unreliable rainfall of the western areas. They are relatively undeveloped, and employment opportunities for large numbers of Africans are quite inadequate. Africans who already live there have to travel to work in the cities or on white-owned farms. The homelands occupy only a little over 10 % of South Africa's area so it seems unlikely that they could accommodate the large numbers of Africans. Also, their existing population of Europeans, coloureds and Asians would have to be moved out.

Racial problems in the United States of America

When Europeans first migrated to North America they found that it was already occupied by American Indians. Then, in the seventeenth and eighteenth centuries, many Africans were brought as slaves to work on the tobacco plantations of Virginia and the Carolinas, and later on the cotton plantations of Alabama and the Mississippi valley. Thus the United States became a multi-racial society.

Most of the American Indians came to be confined to Indian reservations, mostly located in the arid and semi-arid western mountain areas, where the land was generally unproductive.

When slavery was abolished the Negroes of the southern states became poor sharecroppers, and a strict colour bar was observed. In the nineteenth century the majority of American Negroes lived in the area to the east of the Mississippi and to the south of Washington. In the twentieth century however, several million have migrated north, mainly to the cities of north-east USA and the Great Lakes area. They generally moved in search of work and to try to escape from discrimination. However, discrimination also existed in the north-east, so Negroes have tended to congregate in the ghettos which can be seen in almost all large American cities. White immigrants into the USA, such as Italians, have also tended to create ghettos, in which they could live until they had learnt English and had become adapted to the American way of life. But the Negro has tended to remain in his ghettos; he has remained segregated because of his distinctive skin colour. Though living conditions in the ghetto are very poor, the Negro may see it as a refuge from racial discrimination.

Ghettos often began about 1900 as Negro migrants from the south-east built shanty towns near the city centre, but they soon expanded into areas of old housing nearby. As new migrants arrived, population pressure in the ghettos caused them to expand by displacing white Americans from immediately surrounding areas. As white Americans moved to the suburbs in the 1950s to build larger houses with gardens, ghettos were able to expand further. Now the Negro ghetto of a large American city often consists of a wedge-shaped area extending from the city centre almost to the suburbs. Its basic feature is the predominance of Negroes, but it may show variations in affluence ranging from very poor people occupying old, run-down property nearest to the city centre to middle-class people living in pleasant houses nearer the outskirts of the city. Even if a Negro becomes better off he may find it difficult to escape from the ghetto because of the colour of his skin.

2.2 Religions

Five main religions are dominant over the world as a whole. These are Christianity, Islam (the Muslim religion), Buddhism, Hinduism and Confucianism (Fig. 2.1). Christianity is easily the most widespread, being predominant in Europe, the Americas, Australasia and much of southern Africa. Islam predominates in the Middle East and northern Africa. Hinduism, Buddhism and Confucianism are largely confined to southern and eastern Asia.

CHRISTIANITY

Early Development

Christianity originated in Palestine (Israel) in the eastern Mediterranean and soon spread through the Roman Empire in Europe. At an early date it divided into the Roman Catholic Church in western Europe and the Eastern Orthodox Churches in eastern Europe. Much later, Protestant Churches developed, particularly in northern Europe.

Present distribution

After the 'Great Age of Discovery', when Europeans began to migrate overseas in large numbers, they took the Christian religion with them. North America was mainly occupied by northern Europeans who were generally Protestants, except for the French, the Irish and the Poles who were mainly Roman Catholics. Hence, although over most of North America Protestants are more numerous, Roman Catholics predominate in the Canadian province of Quebec, where the French language is spoken, and neighbouring parts of New England in the United States. Similarly, Australia and New Zealand, opened up mainly by northern Europeans, are chiefly Protestant. Most European migrants to Central and South America came from Spain and Portugal. Hence these areas are strongly Roman Catholic in character. In Africa Christianity tends to predominate in areas, largely south of the equator, where Europeans settled, for example in South Africa and on the East African plateau (especially Kenya). In the less developed parts of southern Africa, traditional beliefs are most common. In the USSR the Russian Orthodox Church has a slight lead over Islam. The Muslims are mostly concentrated in the south, near to the Islam-dominated Middle East.

ISLAM

Development

Mohammed, the founder of Islam, lived in Mecca in what is now Saudi Arabia about the 6th century AD. The Islam religion spread rapidly westwards to Spain and North Africa and eastwards towards India, carried by Arab invaders. It was also carried along sea trading routes into the Indian Ocean to reach the coasts of East Africa and south-east Asia.

Present distribution

Islam is now the chief religion of the Old World Arid Belt (Fig. 1.2) which separates the European from the Indian and Chinese population clusters. Over much of this area, at least 90% of the population are Muslims.

In Europe, only Albania and part of Turkey are predominantly Muslim. Islam has also expanded southwards in Africa to become predominant in most of the countries that lie south of the Sahara desert and north of the equator, the chief exceptions being Ethiopia and Ghana, Liberia and others along the coast of the Gulf of Guinea. On the shores of the Indian Ocean, Somalia (in East Africa), Bangladesh, Malaysia and Indonesia are mainly Muslim.

The Loubane mosque in Agadir, Morocco

OTHER MAJOR RELIGIONS

Buddhism

Buddhism originated in northern India. It has since expanded southwards, through India, to Sri Lanka, northwards into Tibet (which is now part of China), and south-eastwards into south-east Asia. Over 80% of the populations of Burma, Cambodia and Thailand are Buddhists.

Hinduism

Hindus are very strongly concentrated in India, but they also exist in considerable numbers in areas to which Indians have emigrated. They form the leading sect in Guyana in South America (Fig. 2.1) and there are considerable numbers in the province of Natal on the east coast of South Africa (see page 22).

Confucianism

Confucianism is largely confined to China. Half of China's population are Confucianists.

The Buddhist Temple of the Dawn, Bangkok, Thailand

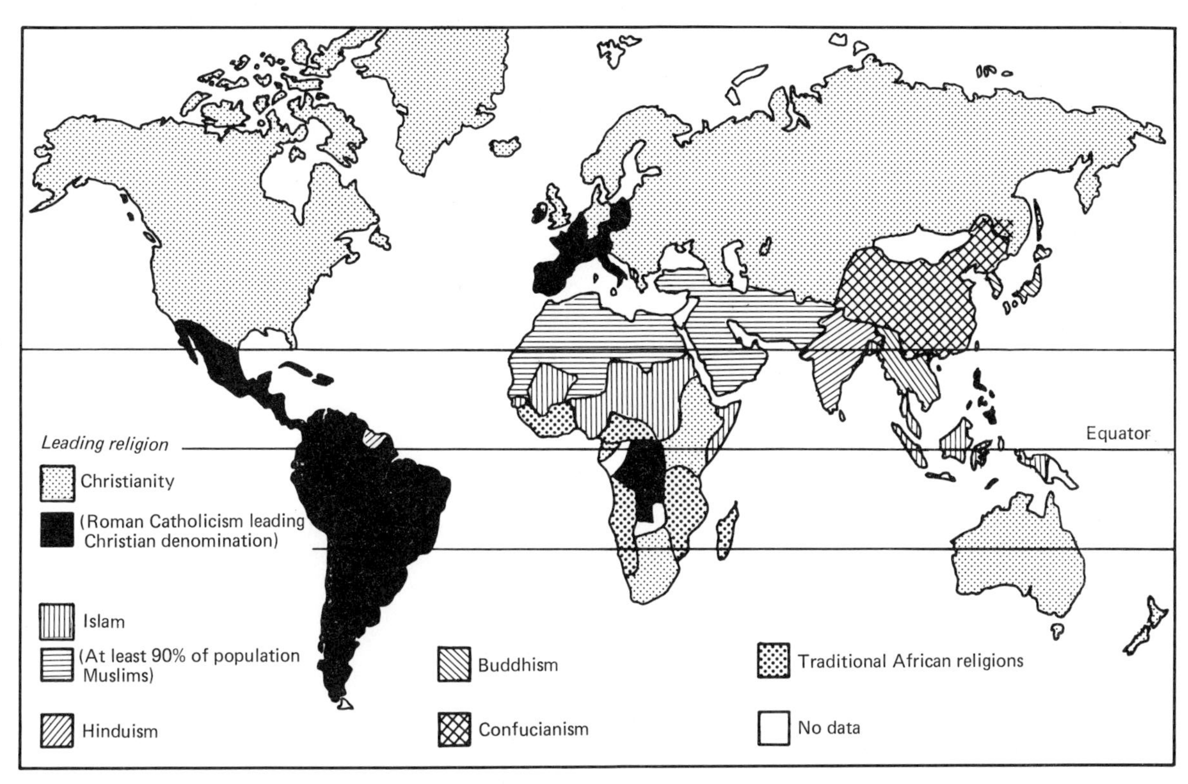

Fig. 2.1 World distributions of major religions

Judaism

Judaism, like Christianity and Islam, originated in the Middle East, but it developed over a thousand years before Christianity. It is widely distributed over the world, with a particular concentration in the state of Israel.

SOCIAL PROBLEMS ARISING FROM RELIGIOUS DIFFERENCES

Many problems have arisen in the past as a result of peoples of different religions occupying the same territory. One of the most serious of these arose through the occupation of the Indian subcontinent by both Hindus and Muslims. The result of this was that the subcontinent was partitioned into two separate countries: India primarily for the Hindus, and Pakistan for the Muslims. Millions of Hindus migrated to India from Pakistan, and many Muslims migrated in the opposite direction. This formed one of the greatest migrations of people that has ever occurred. At the present time there are still a great many social problems which arise from religious differences.

The problem of Northern Ireland

The movement of English and Scottish people into Northern Ireland began well over 400 years ago. The Irish were Roman Catholics, but the immigrants were Protestants. Also the immigrants brought with them the English language and the native Gaelic language declined in importance. At the present time only about 35 % of the population of Northern Ireland are Roman Catholics.

The Protestants and the Roman Catholics tend to occupy different parts of Northern Ireland. Most of the towns tend to be Protestant and the rural areas Roman Catholic, but Protestants predominate in the Belfast area, to the north in Antrim, and along the Lagan valley to the southwest. Roman Catholics are more numerous in the far west and south, near the boundary with the Republic of Ireland, especially in the counties of Tyrone and Fermanagh and in the southern parts of Armagh and County Down.

In Belfast itself Protestants and Roman Catholics tend to occupy separate zones of the city, divided by very sharp boundaries. The main concentration of Roman Catholics forms a belt

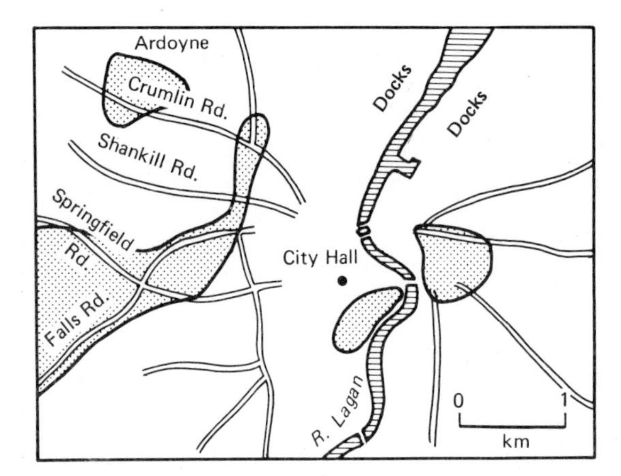

Fig. 2.2 Roman Catholic areas of Belfast

running south-west from the city centre along the Falls Road. A separate smaller concentration is found in the Ardoyne, to the north-west of the city centre, on the Crumlin Road. Between these two Roman Catholic areas a Protestant zone extends along the Shankill Road. Another mainly Protestant area exists in Sandy Row, to the south-east of the Falls Road. Roman Catholics also occupy much of the inner city to the east of the city centre (Fig. 2.2). In the outer residential districts the division between Protestants and Roman Catholics is not so clear.

The religious differences in Northern Ireland are expressed in political disagreements. The Protestants far outnumber the Roman Catholics and they have had an easy parliamentary majority. The Roman Catholics have claimed that they have been the victims of unfair discrimination by the more powerful Protestants. Also, whereas the Protestants favour maintaining close links with the rest of the United Kingdom, the Roman Catholics have tended to support political union with the Republic of Ireland. The conflict which has continued since the late 1960s is a reflection of these religious and political differences.

2.3 Languages

The world distribution of languages (Fig. 2.3) is similar in many ways to the distribution of religions (Fig. 2.1). Most of the world is dominated by the two chief religions, Christianity and Islam, which originated in the Middle East. Only in India and the Far East are there considerable numbers of Hindus, Buddhists and Confucianists, whose religions originated there. Similarly, most of the world is dominated by European or Middle Eastern languages, notably English, French, Spanish, Portuguese and Arabic, but in India and the Far East most people still speak Eastern languages.

NORTH AMERICA

North America has been populated largely by migrants from Europe. The earliest of these were the British and the French, and to some extent Spaniards in the south-west.

Both the United States and Canada are mainly English-speaking, but there are considerable numbers of French-speakers in the Canadian province of Quebec and nearby parts of north-eastern USA. A glance at an atlas map will show several French place-names in this area, such as Trois Rivières and Île d'Orléans. Similarly, in south-western USA one can see Spanish place-names such as San Francisco and San Diego. However, although over a quarter of all Canadians speak French, English is easily the main language of North America, being spoken by immigrant Germans, Poles, Italians and Africans.

CENTRAL AND SOUTH AMERICA

Early immigrants to Central and South America were mainly Spaniards and Portuguese. The Spaniards came to dominate the Pacific coastlands from Mexico and the Panama isthmus, along the line of the Andes to Cape Horn, but also including Argentina and many of the West Indian islands. This area remains Spanish-speaking. The Portuguese on the other hand came to dominate the Atlantic coast from the mouth of the Amazon almost to the River Plate. Hence, the language of Brazil is Portuguese.

The West Indian islands tend to have the language of the European country with which they have had the closest links in the past. Thus, Cuba is Spanish-speaking, Haiti French-speaking and Jamaica English-speaking.

AFRICA

Africa has a great variety of languages, not all of which are of European origin. Swahili, for example, is widely spoken in East Africa. In

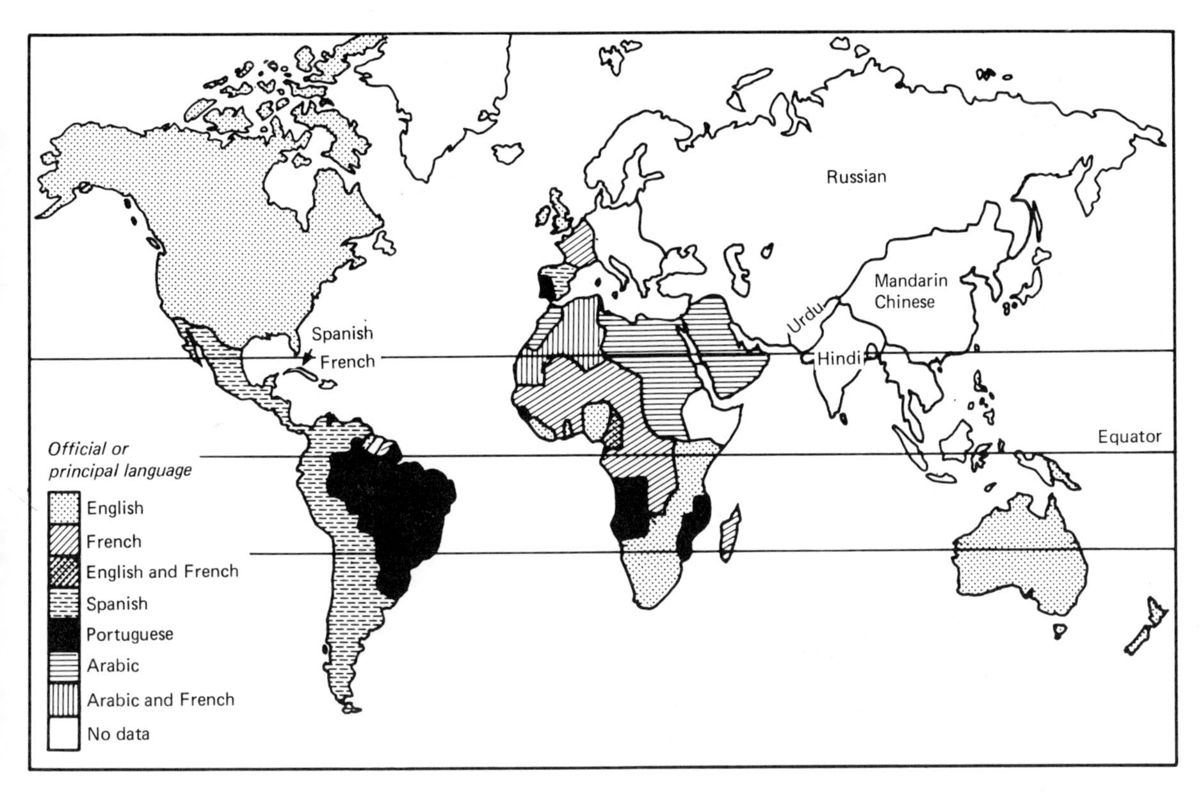

Fig. 2.3 World distribution of major languages

general, however, the pattern of languages tends to reflect the pattern of political influence in the past.

Much of northern Africa is Arabic-speaking, in particular those areas where the Muslim religion is dominant, but French is spoken in countries such as Algeria which have had the closest associations with France. French is also the main language in many parts of central Africa which were once French or Belgian possessions, and also in Madagascar.

Portuguese is the main language of Angola and Mozambique, where Portuguese influence has been greatest.

The English language predominates in South Africa and Zimbabwe, which have received many British immigrants, but many people speak Afrikaans in South Africa. English is also the main language of the Commonwealth countries of East Africa (Kenya, Uganda and Tanzania) and West Africa (Nigeria, Sierra Leone, Ghana and The Gambia) (Fig. 2.3).

PROBLEMS ARISING FROM LANGUAGE DIFFERENCES

The problem of Quebec in Canada

In Canada as a whole only a little over a quarter of the population claim that French is their mother tongue, but in the province of Quebec (Fig. 2.4) this figure rises to over 80 %. The French-speaking population of Quebec has never been absorbed into the Canadian way of life as successfully as other European immigrants such as the Scandinavians and the Dutch. They have tended to keep a separate culture from the rest of Canada by maintaining and developing separate links with France. Many students have gone to study in French universities for example.

Many French Canadians have claimed that they have been discriminated against in various ways by English-speaking Canadians. In 1967, President De Gaulle of France appeared to encourage separation in Quebec by concluding a public speech with the words 'Vive le Québec Libre!'

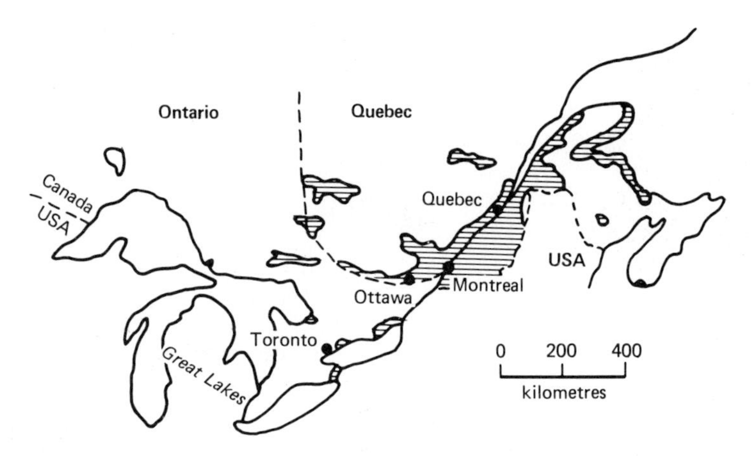

Fig. 2.4 Main concentrations of French Canadians in Eastern Canada

('Long live Free Quebec!'). Suggestions have been made to set up Quebec as a separate state from Canada, but many French Canadians oppose this.

The language problem in Belgium
In Belgium there are two official languages: Flemish Dutch, spoken by the Flemings, and French, spoken by the Walloons. In addition a small proportion of Belgians speak German. Flemish is mainly spoken to the north of the latitude of Brussels (Fig. 2.5) and French mainly to the south. German is spoken mostly near the German border in the east. In Brussels itself both Flemish and French are spoken, but French is dominant. Just over half of Belgium's population speak Flemish.

The main problem is that most of the people who work in managerial or professional jobs tend to be French-speakers. Also, much of higher education in universities and colleges tends to be taught in French. The Flemish-speakers, who are a majority of Belgium's population, have often felt, therefore, that they have been deprived of reasonable opportunities because of the language which they use. The Flemish-speaking population has tended to grow more quickly than the French-speaking population, so that gradually a greater equality of opportunity has been gained.

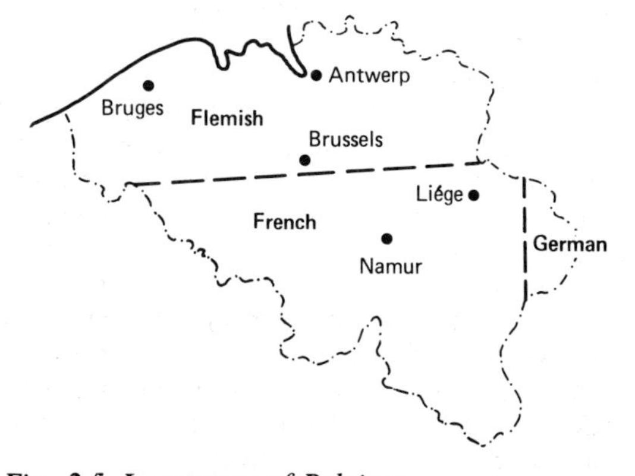

Fig. 2.5 Languages of Belgium

2.4 Political organization

People have always organized themselves into social groups of various sizes. Long ago the social or political unit was the tribe, probably consisting of a number of families linked together by kinship. As time passed however political units tended to become larger, and the nation-state evolved. This is the stage that we have now reached. People feel that they 'belong' to a particular country, and this feeling is expressed in the great interest which is shown, for example, in the Olympic Games or international soccer and rugby matches.

THE EVOLUTION OF THE POLITICAL UNIT

Various kinds of large political units have existed in the past.

Empires
Great empires existed thousands of years ago

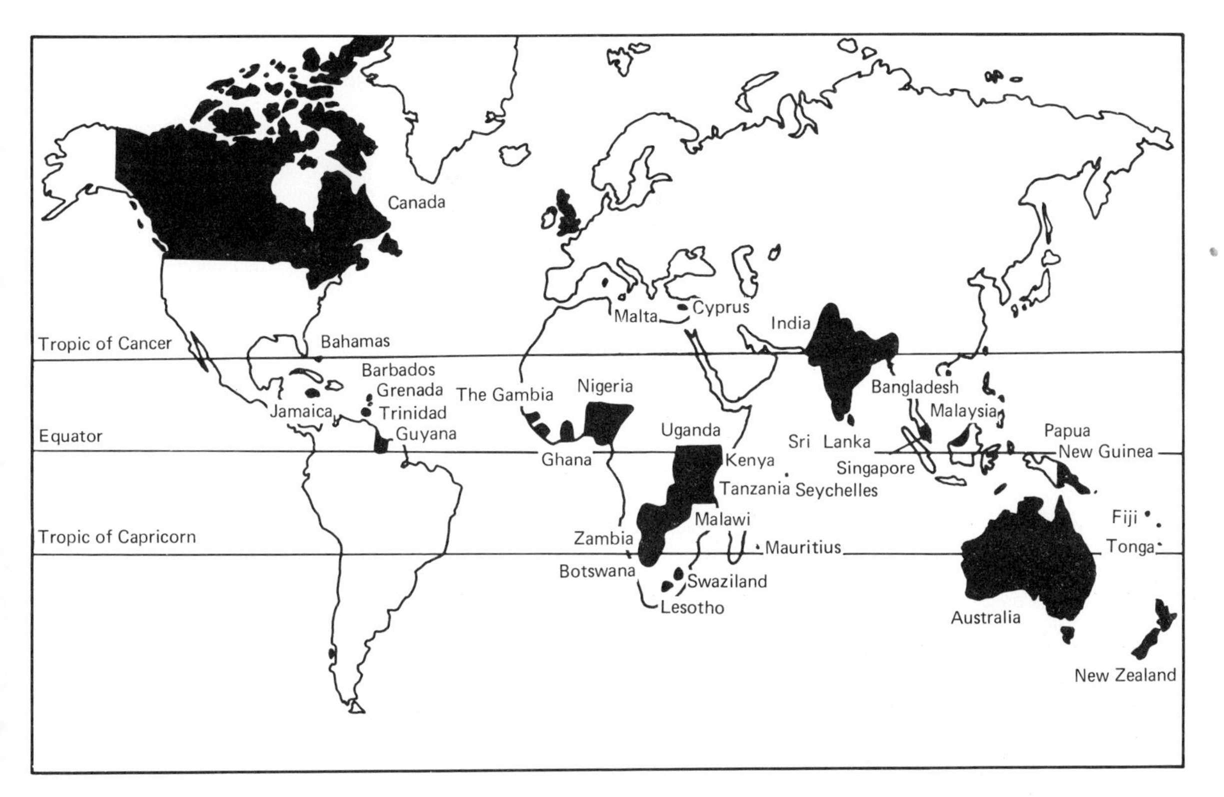

Fig. 2.6 The Commonwealth

when most societies were still at the tribal stage. The Romans for example, by the use of military power, created a great empire surrounding the Mediterranean Sea and extending northwards to Britain and Germany. Other great empires were created by the Egyptians, the Persians and the Chinese.

In the last few hundred years, great empires were created by various European countries, including Britain, France, Belgium and the Netherlands. Such empires linked developed, industrial European countries with underdeveloped countries in Africa, Latin America and the Far East. These modern empires disintegrated in the mid-twentieth century as ideas of democracy and self-determination developed. Thus, many political problems arose as former colonial countries began to govern themselves. Such problems dominate the political geography of Africa at the present day.

Nation states
Nation states were created when people occupying a particular territory and sharing a common history, culture and language discovered a sense of identity and common interest. England and France developed over hundreds of years to become nation states. Others, such as Italy and Germany were created in the nineteenth century. The idea of the nation state is now spreading rapidly from Europe to the rest of the world as former colonies struggle to find a sense of identity.

Supra-national organizations
In the twentieth century there has been a great development of co-operation between nation states to form organizations composed of many different states for either political or economic purposes. The largest of these is the United Nations Organization (UNO) to which belong all the world's nation states except for a few very small ones, mostly in Europe. The aim of the United Nations Organization is to maintain international peace and security and to encourage cooperation.

The former British Empire is now represented

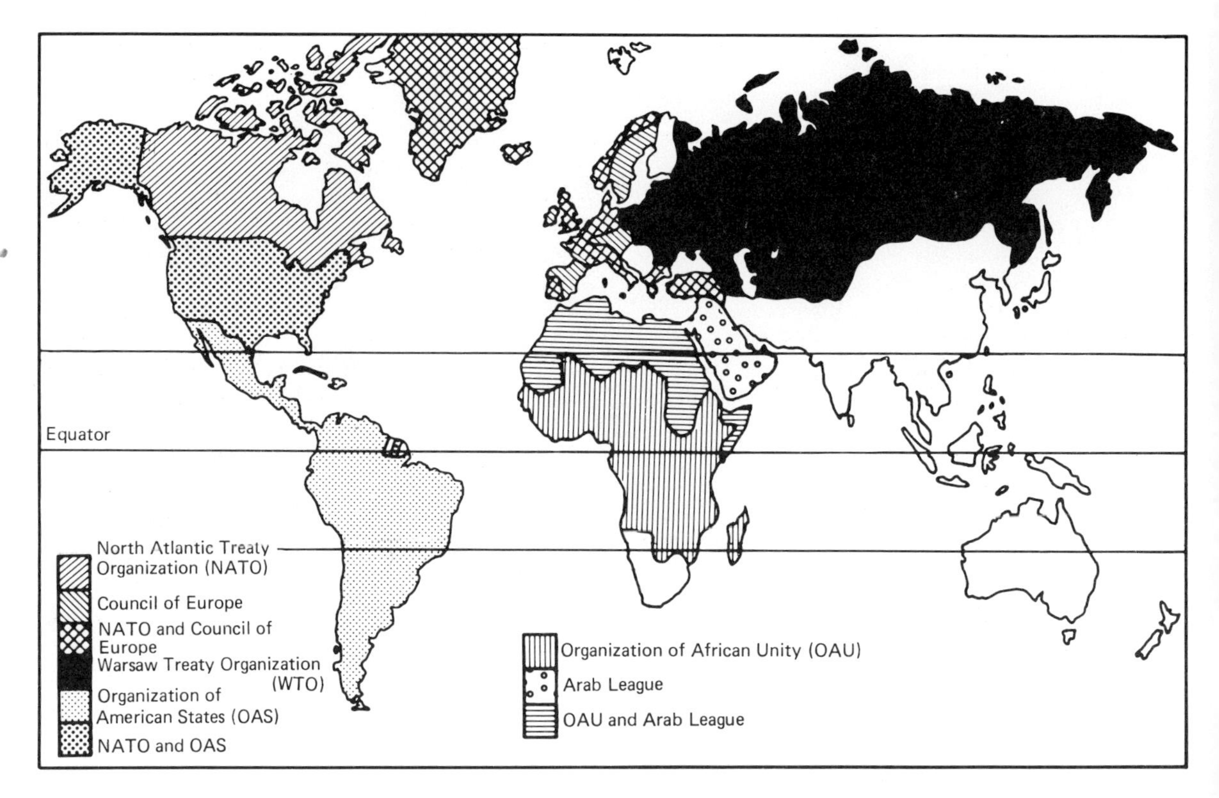

Fig. 2.7 Supra-national political organizations

by The Commonwealth of Nations (Fig. 2.6) which is described as a free association of sovereign, independent nations for the purposes of co-operation and mutual assistance.

The fundamental political division in the world today between the Communist and non-Communist countries is represented by the Warsaw Treaty Organization (WTO) and the North Atlantic Treaty Organization (NATO), both of which exist mainly to provide mutual support if any member country is attacked (Fig. 2.7).

In addition, a number of other organizations exist in various other parts of the world, mainly to defend the sovereignty of their member countries. These include the Organization of American States (OAS) comprising most of the countries of North and South America except for Canada, the Organization of African Unity (OAU), the Council of Europe (to promote European unity) and the Arab League (to promote Arab unity) (Fig. 2.7). One of the objectives of the Organization of African Unity is to eradicate colonialism.

The United Nations headquarters in New York

Other organizations are mainly intended to promote economic growth in their member countries. These include the European Economic Community (EEC) (the United Kingdom, France, West Germany, Italy, Belgium, Luxembourg, the Netherlands, Denmark, Ireland and Greece) and the European Free Trade Association (EFTA) in Europe; the Latin American Economic System (LAES) (the West Indies and Central and South America), the Latin American Free Trade Association (LAFTA) (Mexico and South America), and the Association of South East Asian Nations (ASEAN) (Singapore, Thailand, Malaysia and the East Indian islands). The member countries of the Warsaw Treaty Organization, together with Cuba, form the Council for Mutual Economic Assistance (COMECON). Most of these political organizations were set up very soon after the end of the Second World War. The economic organizations were generally set up later, and they have recently begun to form in the less developed parts of the world.

THE POLITICAL PATTERN OF THE WORLD

Figure 2.7 shows the spatial distribution of the political organizations described above.

The temperate lands of the northern hemisphere are generally divided between the countries of the Warsaw Treaty Organization in eastern Europe and the USSR and those of the NATO countries generally around the coasts of the North Atlantic ocean, but also including Italy, Greece and Turkey in the Mediterranean. These two groups of countries are separated in places by countries such as Finland, Austria and Yugoslavia which belong to neither of these rival organizations.

In the Middle East and northern Africa are the states of the Arab League. These generally coincide with the countries where Arabic is the main language (Fig. 2.3) and also with those in which Islam is the main religion (Fig. 2.1), but the mainly Muslim countries of Turkey, Iran, Afghanistan and Pakistan are not members of the Arab League. Turkey appears to look more towards Europe, being a member of both NATO and the Council of Europe. Somalia in eastern Africa, whose main religion is Islam, is also a member of the Arab League, although Arabic is not its principal language.

All the African countries except the Christian, English-speaking ones of the far south, chiefly the Union of South Africa and Zimbabwe, are members of the Organization of African Unity.

The Commonwealth includes a great variety of countries which differ greatly from one another. Some, such as Canada, Australia and New Zealand, are inhabited mainly by people of British descent. South Africa and Zimbabwe are no longer members of the Commonwealth. Other Commonwealth countries include former colonial territories in Africa (e.g. Nigeria, Ghana and Kenya), India and Bangladesh, and a large number of very small islands in the Pacific Ocean (e.g. Fiji and Tonga), the Indian Ocean (e.g. Mauritius and the Seychelles), the West Indies (e.g. Barbados and Jamaica) and the Mediterranean (Malta and Cyprus). These have a very wide distribution over the world (Fig. 2.6) in contrast to NATO, WTO, Arab League and OAU countries, which are localized in particular areas.

The Parliament building, Budapest, Hungary, on Constitution Day

POLITICAL PROBLEMS

Political problems in Africa

Some of the world's most serious political problems exist in the developing countries of Africa to the south of the Sahara Desert. These countries are shown in Figure 2.8. Almost all these countries have severe population problems. In most of them the birth rate is well over 40 per 1000 per annum and in several it is over 50 per 1000. Declining death rates are giving rates of population increase of about 3% per annum. So far, however, population densities are relatively low except in parts of East Africa and along the Gulf of Guinea coastlands (Fig. 2.8). Living standards are generally low (Fig. 2.9) and most of these countries are comparable with India and Pakistan in this respect.

Figure 2.10 shows the total populations of these countries and their location in relation to one another. Nigeria has a greater population than any two other countries added together. In East Africa, Uganda, and particularly Rwanda and Burundi, have large populations in comparison with their size. If political influence is proportional to population numbers it is easy to see from this diagram which of the countries are likely to have the greatest influence.

Many African political problems result from the influence of European countries between the late nineteenth century and the 1950s. Europeans divided Africa into colonies without paying much attention to the characteristics and the traditional loyalties of the various groups of Africans. The result was that, when they became independent in the mid-twentieth century, these new African

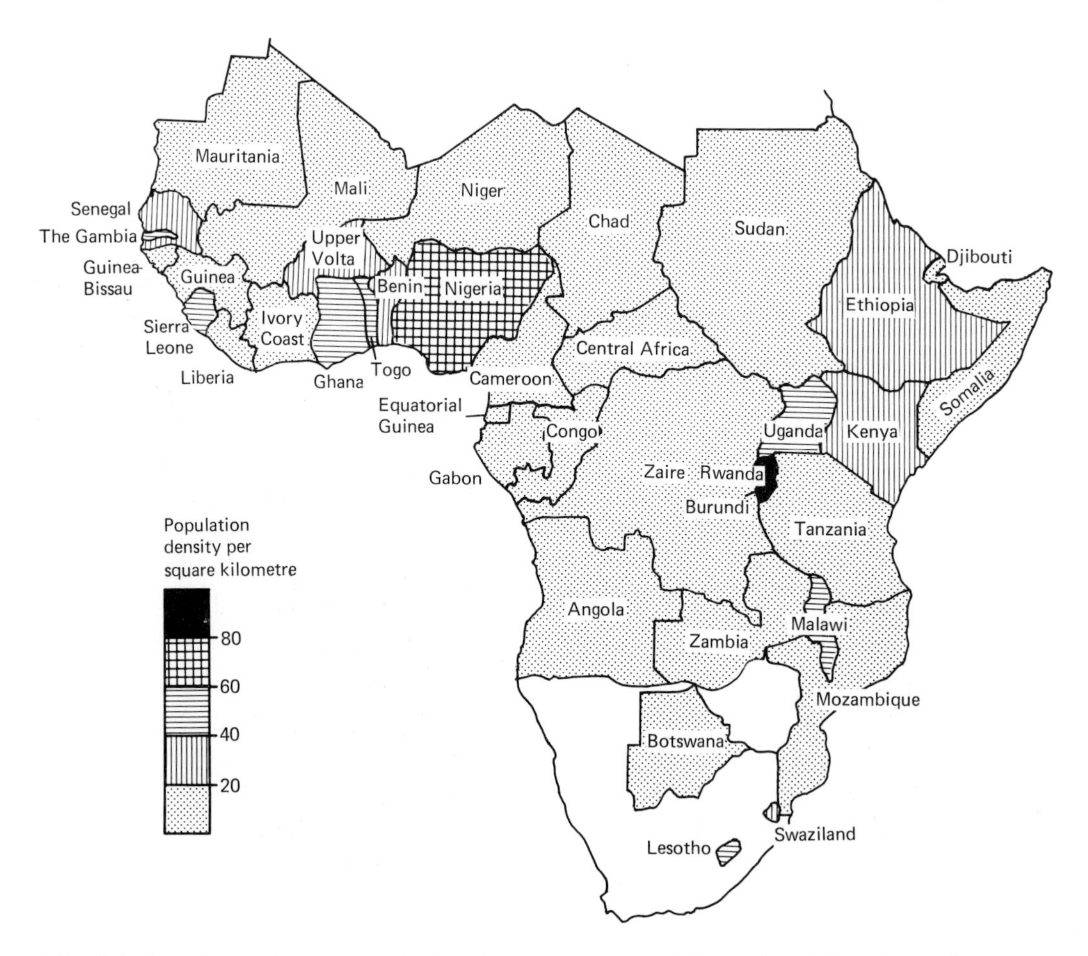

Fig. 2.8 Population densities in the developing countries of Central and Southern Africa

states contained many conflicting groups of people. It has been very difficult to build up a national loyalty among peoples who felt a loyalty only to their particular ethnic or tribal group. Thus internal conflicts have frequently broken out. A particularly serious example occurred in the late 1960s when Biafra tried to break away from Nigeria and a civil war took place which caused great suffering. One of the aims of the Organization of African Unity is to build up loyalty to these new nations within the existing framework of political boundaries.

Another associated political problem is that of absorbing non-African peoples, such as Indians, into these new nation-states.

Political problems concerning the sea

Up to the middle of the twentieth century the sea was regarded mainly as a medium for transport and a source of fish. For hundreds of years coastal states had claimed various widths of adjacent sea as their sovereign territory (territorial sea) in early days for purposes of defence. Many of these states claimed a three-mile (5 km) wide zone of territorial sea, but this was not a generally accepted principle.

Since the 1940s the situation has changed considerably because it has come to be realized that the sea is an important source of natural resources, and the technology has been developed to exploit them. These include the following.

Large deposits of petroleum and natural gas exist in the rocks which underlie the continental shelf, particularly in north-west Europe, the Mediterranean, eastern North America and the East and West Indies. Hence, it has been necessary to divide the North Sea between Britain, Norway,

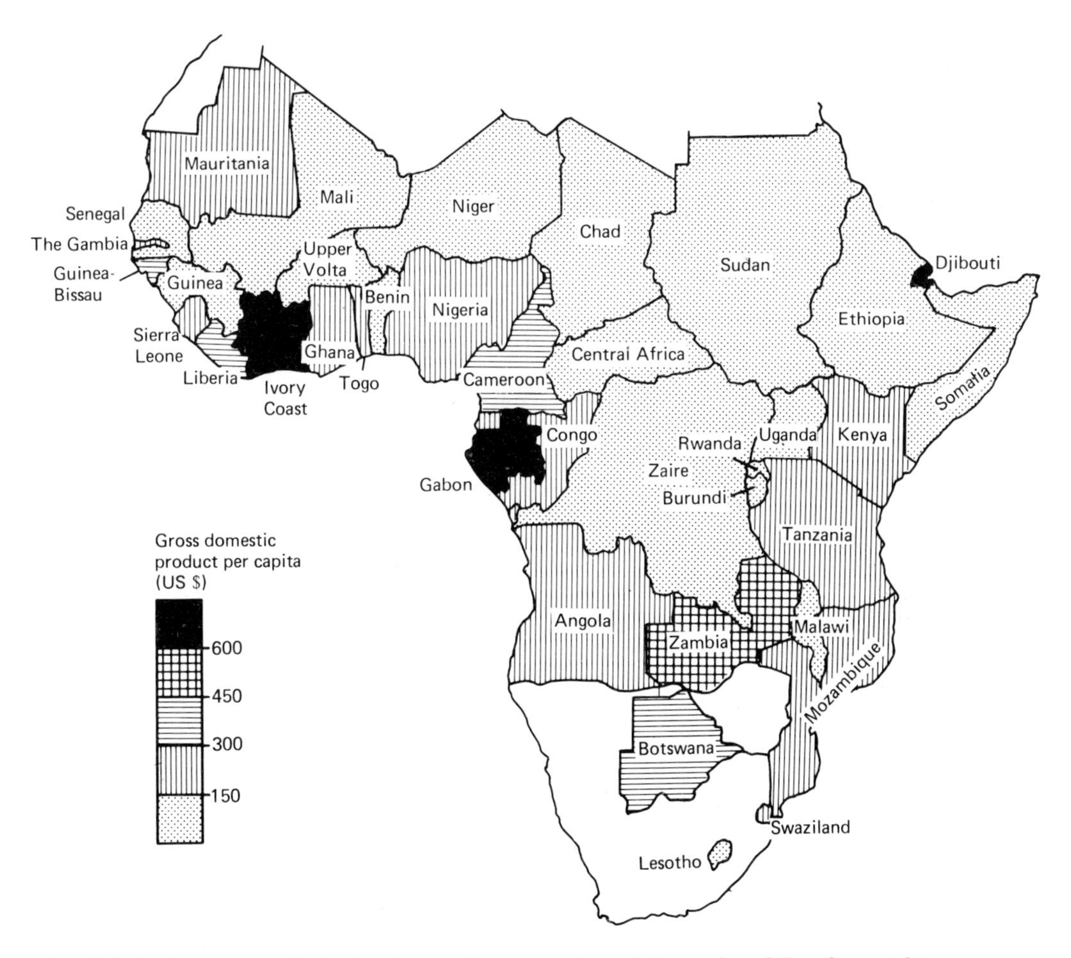

Fig. 2.9 Living standards in the developing countries of Central and Southern Africa

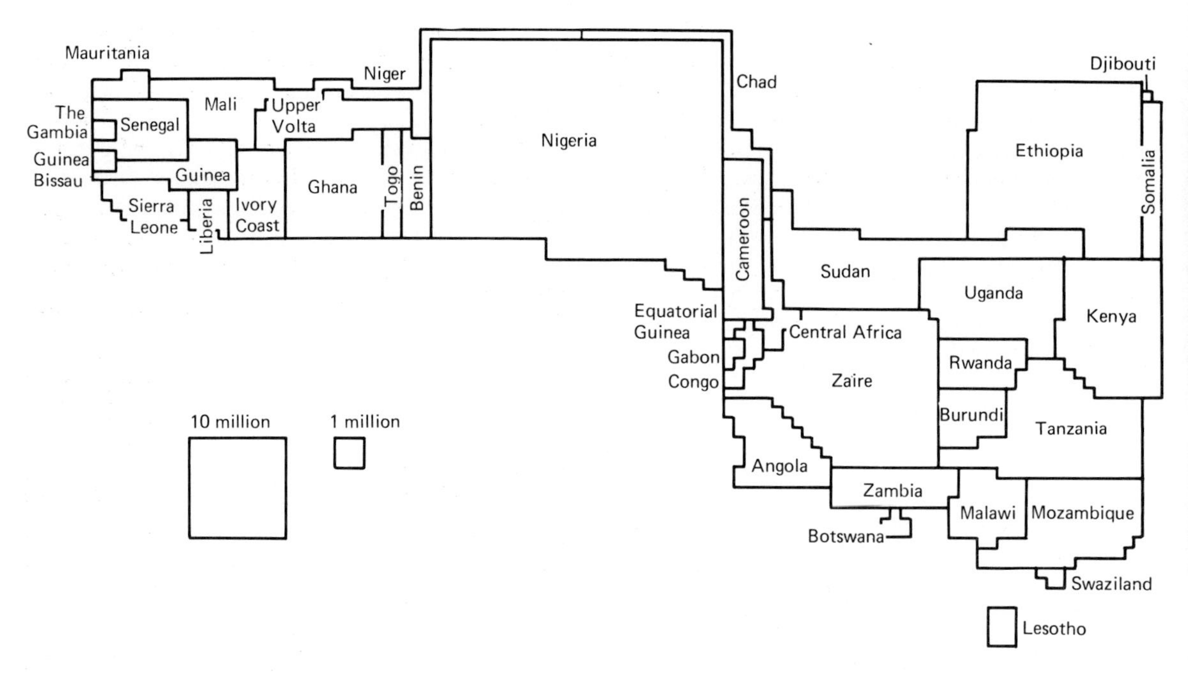

Fig. 2.10 Population totals of the developing countries of Central and Southern Africa

the Netherlands and other coastal states in order to determine the ownership of its petroleum and gas reserves.

Demand for fish has increased and various countries have laid claim to control of certain fishing grounds. In the 1970s much friction existed between Britain and Iceland in this connection. In 1972, Iceland extended its fishing zone from 12 to 50 miles (19 km to 80 km) offshore. Britain objected to this and the outcome was that British trawlers were allowed to catch a specified amount of fish in particular parts of the zone lying between 12 and 50 miles offshore. In 1975, Iceland announced an extension of its fishing limits to 200 miles (320 km) offshore. This led to another 'cod war' in which Royal Navy vessels were sent to protect British fishing boats within this 320 km zone.

The importance of the mineral deposits of the sea bed has been realized. Manganese nodules, formed by precipitation from sea water, are widely distributed over the world's seas, and they contain nickel, copper and cobalt in addition to manganese.

Thus a problem exists as to how these resources are to be shared between the countries of the world, and whether they should become the property mainly of those countries that happen to have long coastlines facing wide expanses of sea. There are 30 completely landlocked countries in the world and over half of these are very poor, underdeveloped countries in Africa and Asia, such as Mali, Burundi, Rwanda, Afghanistan and Laos. Many of the richer countries of the world, such as the United Kingdom, Canada, the USA and Australia have long coastlines with great areas of sea fronting them. There appears to be a danger therefore that exploitation of the sea's resources will increase the existing differences in wealth between the advanced countries of the world and the underdeveloped countries.

Exercises

1. Explain the extent to which the present world distribution of:
(*a*) racial types
(*b*) languages
(*c*) religions
is the result of the migration of population.

2. Discuss the extent to which
(*a*) the world distribution of English speaking peoples corresponds to the world distribution of peoples who enjoy high living standards.
(*b*) the members of supra-national political and economic organizations are grouped according to common characteristics of race, language and/or religion.

3. Describe and explain the similarities and contrasts between the social/political problems of
(*a*) southern Africa and the USA
(*b*) Quebec and Northern Ireland.

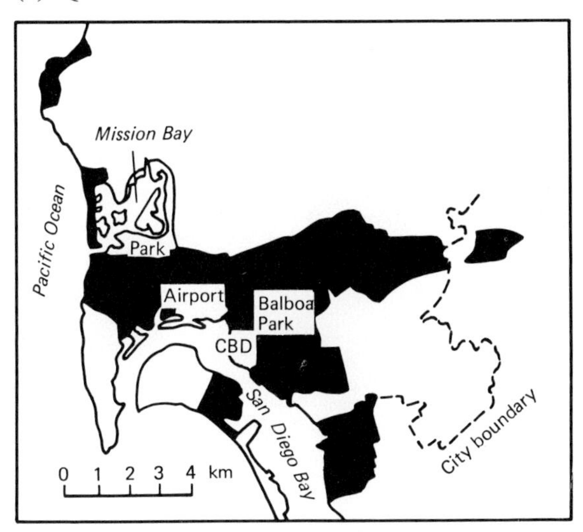

Fig. 2.11 San Diego: built-up area in 1940

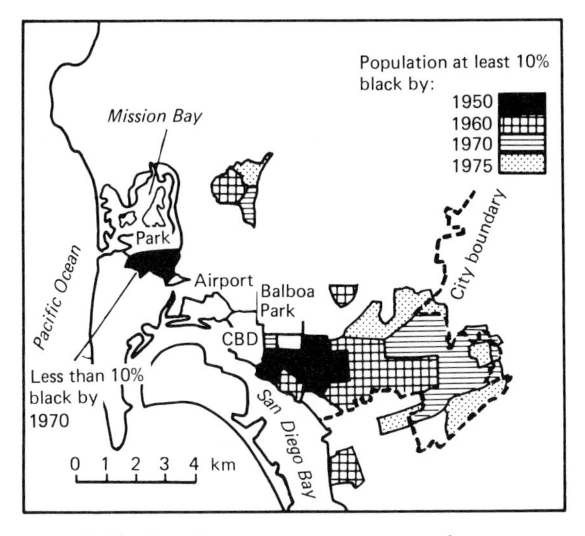

Fig. 2.13 San Diego: census tracts whose population was at least 10% black by the dates given

4. Using the information provided in Figures 2.11–2.14, describe and comment on the spatial evolution of San Diego's black ghetto between 1950 and 1975. To what extent does San Diego's ghetto conform to the generally accepted concept of a ghetto as an area of social deprivation?

Figures 2.11–2.14 are adapted from Ford, L. and Griffin, E.: 'The Ghettoisation of Paradise' (Geog. Rev., vol. 68, April 1979) with the permission of the American Geographical Society

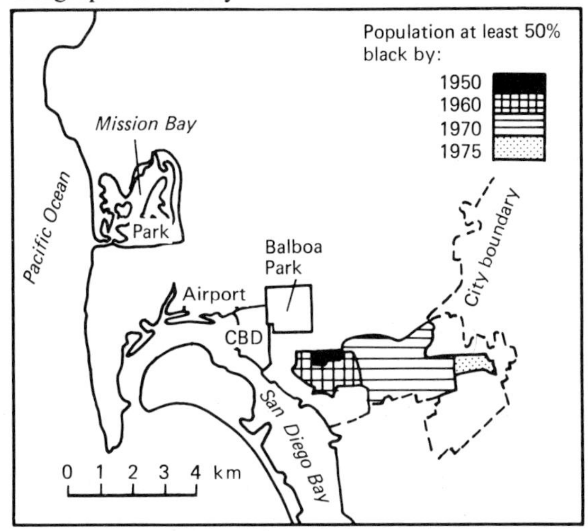

Fig. 2.12 San Diego: census tracts whose population was at least 50% black by the dates given

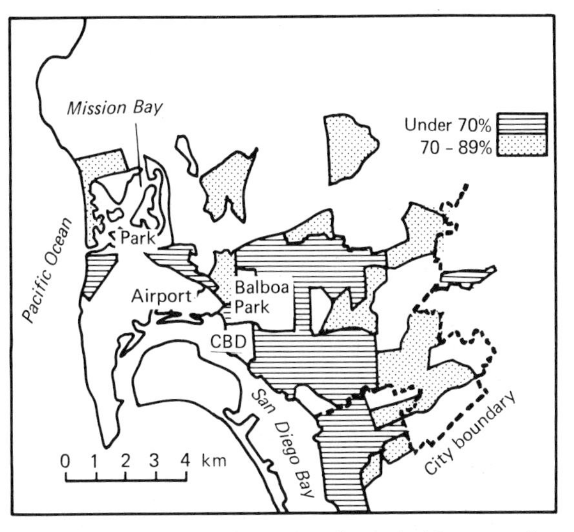

Fig. 2.14 San Diego: median household income by census tract as a percentage of the median income for the whole of the San Diego urbanized area (1975)

3 Economic development

3.1 Population and economic development

THE NATURE OF ECONOMIC DEVELOPMENT

Studies of the history of present-day advanced countries suggest that they have developed economically through a number of consecutive stages.

The traditional (pre-industrial) society
At this early stage people are mostly engaged in farming, as a way of life rather than a commercial enterprise. Power is derived mostly from animals and human strength, though some use may be made of the wind (for boats) and simple water power.

The economy is a subsistence one. Enough food is produced by growing crops and rearing animals to feed people and animals, so that more food is produced.

Social groups are quite small and loyalty is felt chiefly to the local village or tribe. There is little sense of belonging to a nation. Mental attitudes are strongly influenced by tradition.

The transition to an industrial society
Traditional attitudes gradually change and people begin to think that economic development is desirable in order to improve their living standards

Primitive hand industry in Bahrain

and their national political power. This tended to happen in Europe in the eighteenth and nineteenth centuries after the rest of the world had been opened up for trade. Industries could develop and send exports to markets all over the world.

Industrialization really began on a considerable scale in Britain round about 1800. It then spread through parts of Europe and North America during the nineteenth century. In the twentieth century European ideas of industrialization have spread to Japan and other parts of the Far East, Australia and South America.

The industrial society

Often, early industrial development tended to be dominated by a few industries only, such as the textile and iron and steel industries in Britain. These however led to the development of a much wider range of industries as industrial technology developed. New science-based industries, such as chemicals and electronics appeared in the twentieth century. Also automation of manufacturing developed, thus reducing the need for a large industrial labour force.

Living standards have become very high in advanced industrial countries and there is a great demand for expensive manufactures such as motor cars and television sets. However, demand has also increased for entertainment, including holidays and foreign travel, banking and other financial services. Thus a great increase has occurred in employment in service industry which provides such facilities.

Thus it appears that, as countries have developed from subsistence economies to 'affluent societies' the major type of economic activity has changed successively from agriculture to manufacturing industry and then to service industry. This does not necessarily mean, however, that present-day traditional societies will develop in exactly the same way.

The Ginza shopping area of Tokyo

RELATIONSHIPS BETWEEN LIVING STANDARDS AND TYPES OF EMPLOYMENT

The relative decline of employment in agriculture over the world

Figure 3.1 shows how employment in agriculture has declined relative to manufacturing and service industry since the early nineteenth century. Although agricultural employment has suffered a relative decline, actual numbers of people working in agriculture, taking the world as a whole, have increased as the world's population has increased.

It is clear from Figure 3.1 that the relative decline in agricultural employment was a process which began in Europe and then spread at different rates to the rest of the world. By 1850, agriculture employed less than half of the total labour force only in parts of Western Europe. This was the area (Britain, France, Belgium and the Netherlands) where manufacturing industry and trade developed earlier than anywhere else. By 1900, the trend had extended eastwards and north-eastwards in Europe and also into many of the lands overseas which had been settled by emigrant Europeans, e.g. North America, Australia, New Zealand, Argentina, Chile and Uruguay.

By 1950 agricultural employment had fallen to less than half of the labour force in Italy and the Iberian Peninsula (in Europe) and also in Japan, the USSR and South Africa, as these areas became industrialized. By 1970 most of the remainder of North and South America, the Middle East, and even isolated parts of Africa had followed the same trend. Most of the predominantly agricultural countries were now restricted to Africa and southern and eastern Asia where living standards were generally lowest, but some still existed in south-east Europe (Romania and Yugoslavia).

Employment structure and living standards

It is usual to divide types of employment into three main groups. These are as follows.

Primary occupations include agriculture, forestry, hunting and fishing, together with mining and quarrying. These occupations are concerned

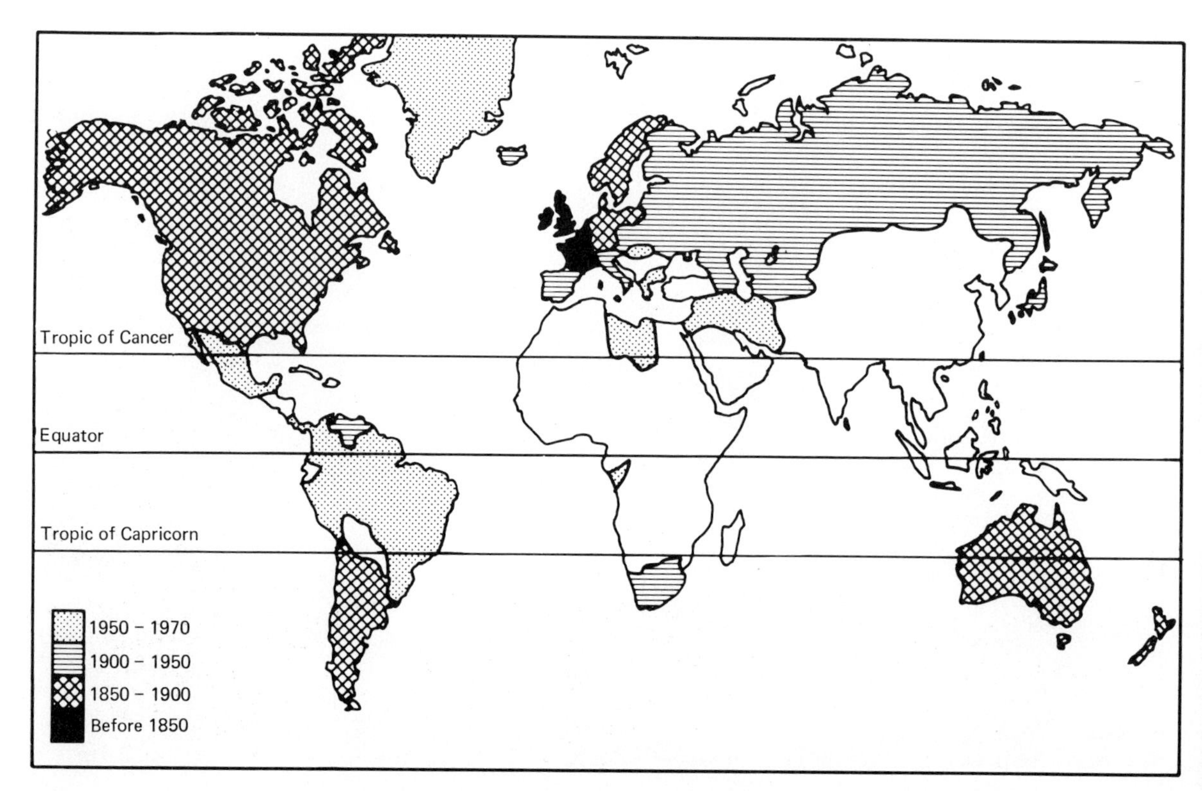

Fig. 3.1 Dates at which agricultural labour force had fallen to 50% of total labour force

with obtaining materials such as crops, timber, fish, coal and mineral ores *directly from nature*. Of these, agriculture employs easily the largest number of people. We can see from Figure 3.1 that employment in agriculture varies greatly over the world in modern times. It is highest in the underdeveloped countries of Africa and south-east Asia and lowest in the developed countries of Europe and North America. This applies also to primary occupations in general. In some under-developed countries they employ over 70 % of the labour force, mostly in agriculture. It is rare for more than 2 or 3 % of the labour force to be employed in mining even in countries such as Chile (copper) and Sweden (iron ore) which are well known for their mining products. In advanced countries employment in primary occupations falls to a very low level indeed (5 % in the United Kingdom and Belgium). Even in countries such as Denmark and Canada, which are often thought to specialize in agriculture, employment in primary occupations as a whole is still under 10 % of the labour force.

Secondary occupations are those in which raw materials obtained directly from nature are processed into a form which makes them suitable to be consumed or used by man. Wheat, for instance, is turned into flour and then into bread, timber into furniture, iron and other ores into metals and eventually machines. In most countries less than 50 % of the labour force are employed in these occupations. The percentage is lowest in under-developed countries, often being 10 % or less. It rises to about 40 %, especially in European countries, but is rather less than this in the USA.

Tertiary occupations are concerned with the provision of various types of services to people. Such services include commerce (mainly the wholesale and retail trades), the provision of transport and communications (e.g. the telephone

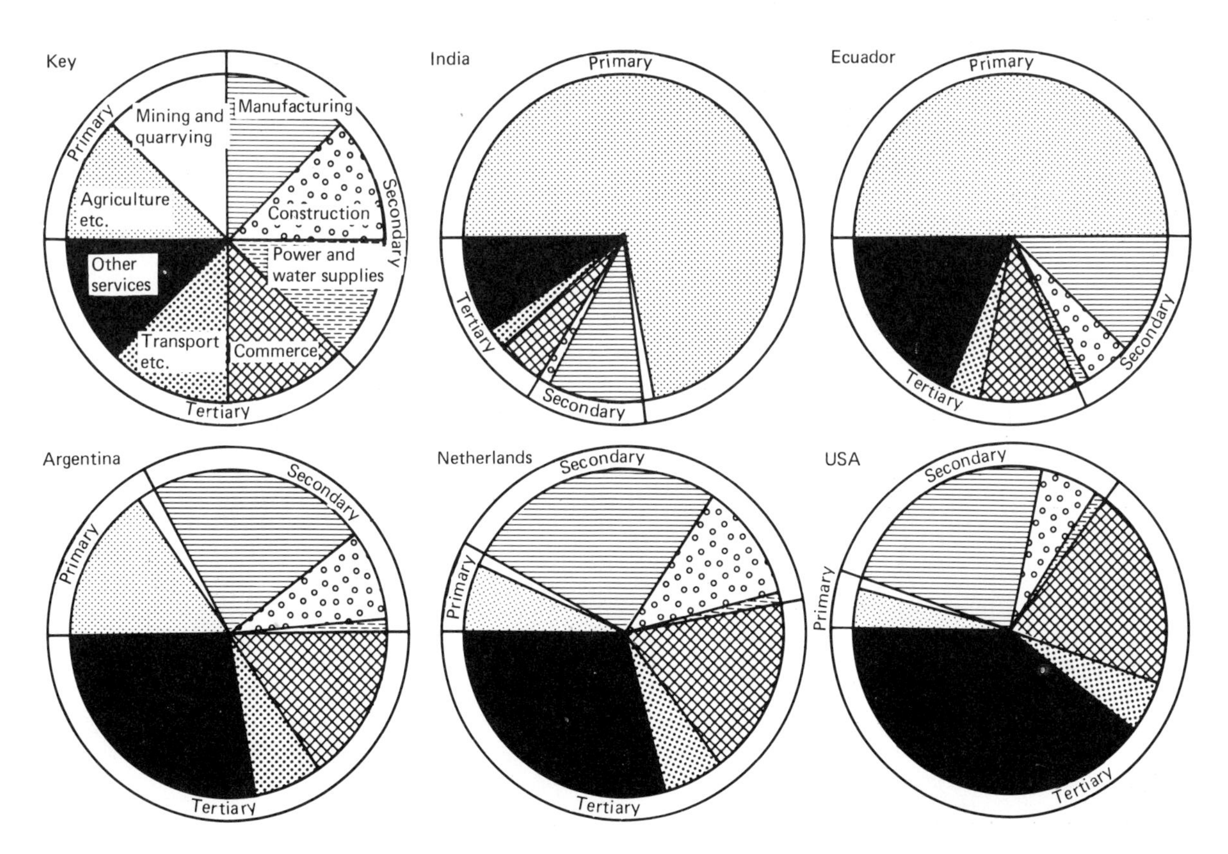

Fig. 3.2 Types of employment in selected countries

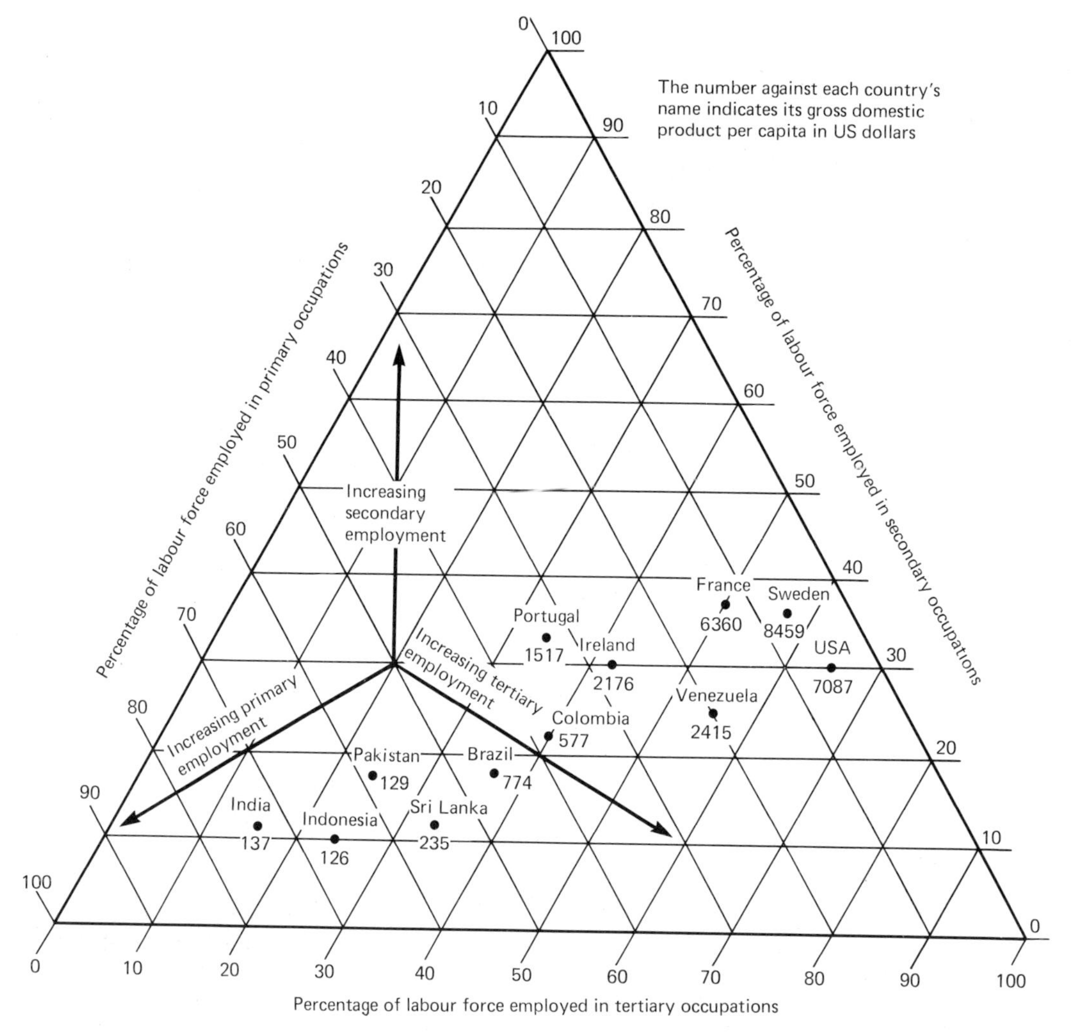

Fig. 3.3 Relationship between type of employment and living standard

and postal services) and various services supplied to industry or to people such as education, entertainment, medical attention and legal services. The Civil Service is also included. These occupations usually employ at least half of the labour force in European countries and Japan. In Canada and the USA well over 60 % of the labour force are employed in this way. Tertiary employment is also relatively high, comprising about a quarter of the labour force, even in underdeveloped countries. In such countries tertiary occupations usually employ double the number who are employed in secondary occupations.

In recent years a tendency has developed to recognize a group of 'quaternary' occupations which are concerned with the provision of tertiary services of very high quality, such as highly specialized knowledge.

The differences in employment structure between five countries which have reached different stages of development are illustrated in Figure 3.2. The five countries range from India, an underdeveloped country, to the USA, which has reached a high level of development. Through this sequence of countries:

(*a*) agricultural employment declines steadily, the proportion of the labour force so employed in

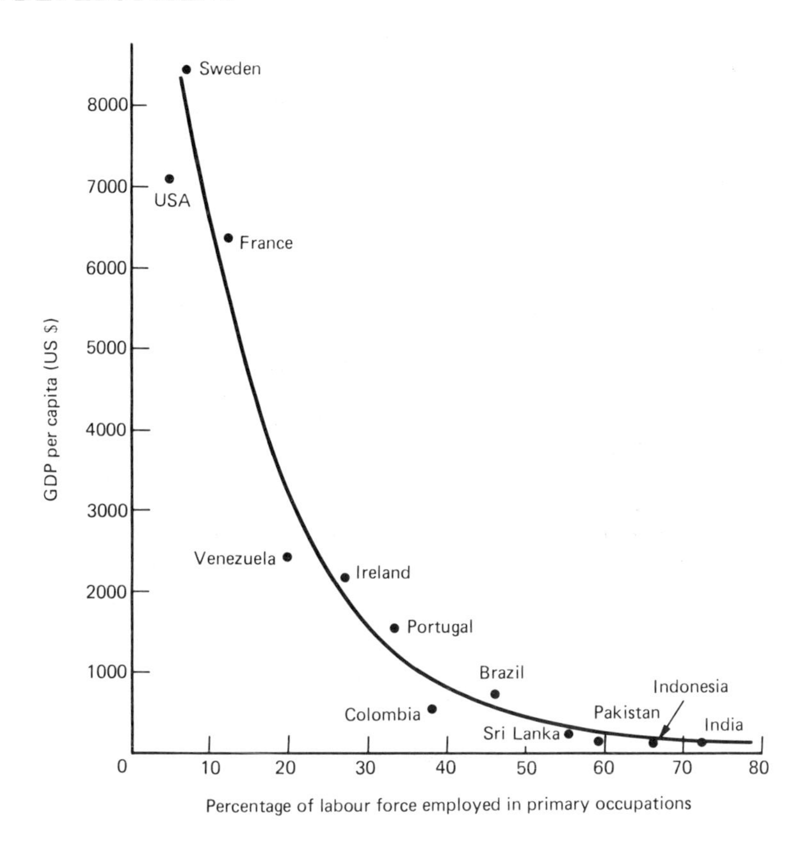

Fig. 3.4 Relationship between employment in primary occupations and living standard

the USA being less than one-fifteenth of that in India;

(*b*) manufacturing employment increases steadily from India through Ecuador and Argentina, to the Netherlands, but then declines slightly in the USA;

(*c*) employment in commerce increases steadily, the proportion employed in commerce in the USA being four times that in India;

(*d*) employment in 'other services' increases steadily, the percentage so employed in the USA being more than four times that in India.

Employment structure is closely related to standards of living. This can be neatly illustrated by means of a triangular graph as in Figure 3.3. This technique permits us to show the percentages of the labour force employed in primary, secondary and tertiary occupations all on the same graph. A selection of twelve countries is marked on the graph, ranging from those with very low values of GDP per capita like India and Indonesia to those with very high values such as Sweden and the USA. The GDP per capita is an approximate measure of the standard of living. The points on the graph form a rough line almost parallel to the arrow showing an increase in primary employment. Thus, it is clear that living standards are related to employment in primary occupations. Sweden and the USA have under 10% of their labour force in primary occupations whereas India has over 70%. The graph (Fig. 3.4) supports this observation. The smooth curve formed by these 12 countries on this graph indicates a very close correlation.

RELATIONSHIPS BETWEEN LIVING STANDARDS AND POPULATION DENSITY

Overpopulation, underpopulation and optimum population

It is possible that there is also a relationship

between population density and standards of living, and that if a country has few people living standards can be higher. Let us find out if this is the case.

The graph in Figure 3.5 shows the relationship between GDP per capita and population density for a sample of 18 countries. Certainly Bangladesh, with the lowest living standard of all the countries shown on the graph, has the highest population density but, apart from this, no relationship can be seen. There are densely populated countries with high living standards (West Germany, Belgium and the Netherlands) and relatively densely populated countries with low living standards (Jamaica, India and Sri Lanka). Sparsely populated countries include Norway, with very high living standards and Chad and Burma, which are almost as poor as Bangladesh.

It cannot even be argued that living standards are related to the natural resources that countries possess. Libya, of course, owes its high GDP per capita to its rich oil reserves, but wealthy countries

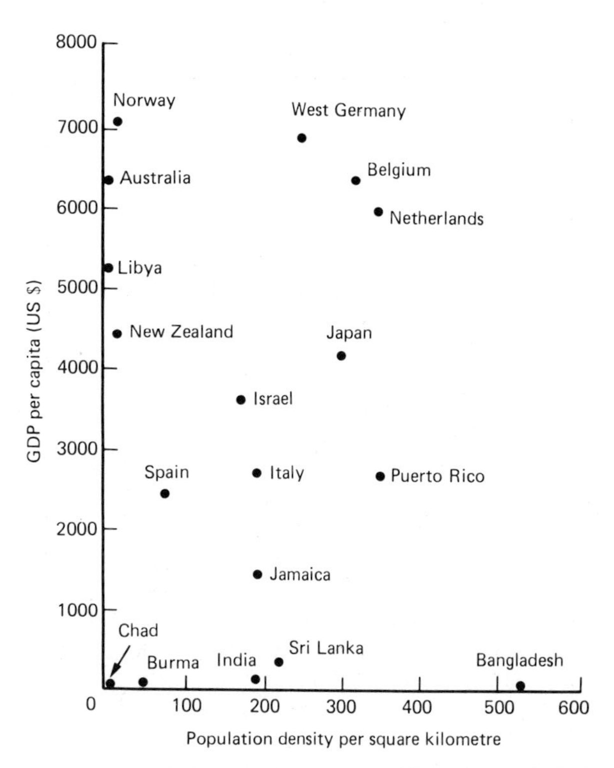

Fig. 3.5 Relationship between population density and living standard

such as Norway, the Netherlands and Japan, have relatively few natural resources. If there is a relationship between population density and living standards it is clearly not a simple one.

The terms 'overpopulation', 'underpopulation' and 'optimum population' refer to the relationships between three variables as follows.

The population total or density is a country's labour supply as well as its consumers. If labour productivity (output per person) is high, then consumption can also be at a high level, and there will be high living standards.

The level of technology is the knowledge and the capital equipment (such as transport networks, irrigation systems, etc.) possessed by the population. With a high level of technology the population will be more productive, so living standards can be higher.

Natural resources include the relief of the land surface, the rivers, the climate, the vegetation, the soils and minerals possessed by a country. These do not determine the standard of living. Living standards depend upon the success with which the population applies its technology to these resources.

The graph in Figure 3.6(*a*) shows how we would expect living standards to change if the population of a country were to increase, but the natural resources and the level of technology were to remain unchanged. As population grows at first from a very low level, productivity is likely to increase as the natural resources are used more fully and division of labour becomes possible. Eventually however productivity would reach a maximum level. If population increases beyond this, with the level of technology unchanged, productivity would begin to decrease and living standards to fall.

The highest point of the productivity curve in Figure 3.6(*a*) represents the *optimum population.* To the left of this the country is said to be *underpopulated.* An increase in population would improve living standards. To the right of the optimum population the country would be described as *overpopulated.* A reduction in population would have the effect of raising living standards.

The level of the optimum population will vary from country to country because natural resources and levels of technology vary from country to country. A density of population which represents underpopulation in one country could represent overpopulation in another and could be the optimum population of another. Just because a country is densely or sparsely populated does not mean that it is overpopulated or underpopulated. A population density of only a few people per square kilometre could represent overpopulation in a desert country unless they were equipped with a high level of technology. This is illustrated in Figure 3.6(*b*) (country 2).

A country's optimum population may change as time passes and technology develops. Improved technology can increase productivity and permit higher living standards to be enjoyed by a greater number of people. This of course has happened in Britain and other countries in the last 200 years. This is illustrated in Figure 3.6(*c*).

It is easier to define an overpopulated country than an underpopulated one. Both could have very low living standards, but the overpopulated country is more likely to have unemployment or underemployment and a considerable amount of outward migration. Inward migration is more likely to occur with an underpopulated country.

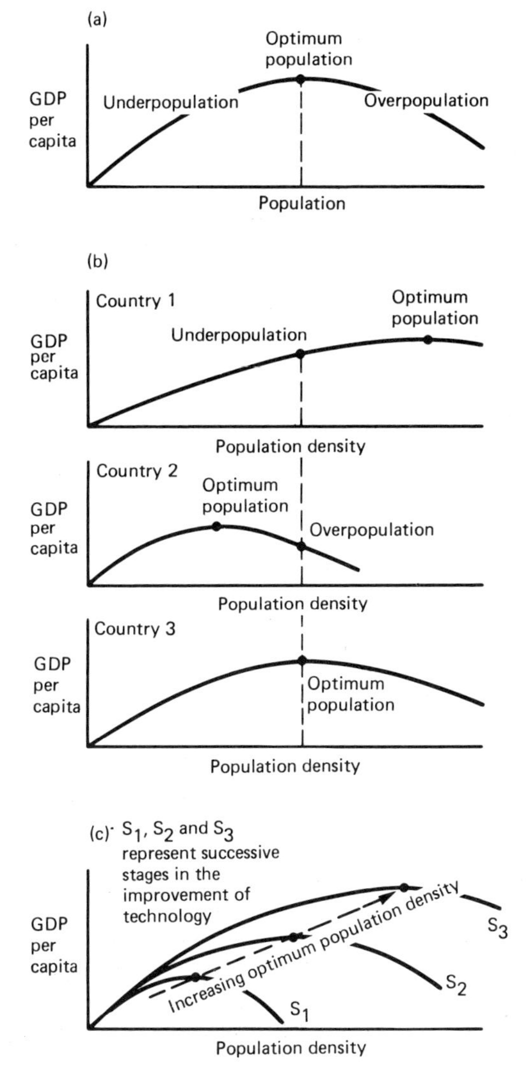

Fig. 3.6 Optimum population, overpopulation and underpopulation

Bangladesh: an example of an overpopulated country

Bangladesh (Fig. 3.7) is almost certainly overpopulated because of a combination of

(*a*) a large, rapidly increasing population;
(*b*) relative poverty in natural resources;
(*c*) a low level of technology which prevents the population from making the best possible use of the natural resources.

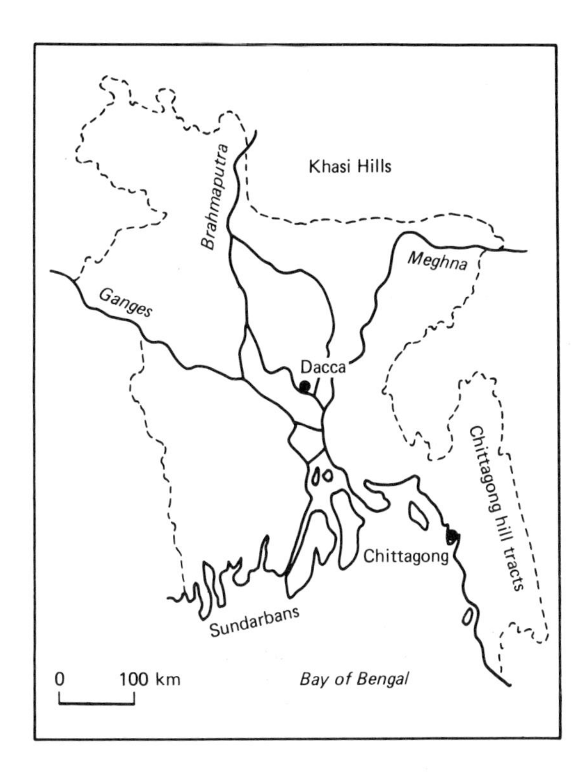

Fig. 3.7 Bangladesh

In the early twentieth century, Bangladesh (then East Bengal, and part of India) had a population of 20 million, and was able to produce a food surplus. Now its population is over 75 million in an area a little smaller than that of England and Wales. The population density is over 500 per square kilometre, over half as great again as that of England and Wales, despite the fact that Bangladesh is a mainly agricultural country. It is the most densely populated country in the world apart from a few very small countries such as Hongkong, Macao, and Monaco, which really only consist of a single city.

Bangladesh consists largely of the delta and the alluvial plains of the lower courses of the Ganga (Ganges) and Brahmaputra rivers (Fig. 3.7). Much of this level land is liable to flooding. The tropical monsoon climate provides enough warmth for crops to grow at all seasons, but the rainfall comes mostly during the summer, three-quarters of the annual total falling between June and September. Irrigation is therefore needed to produce crops during the dry winter. This summer concentration of rainfall also tends to cause disastrous flooding every few years, through which many lives are lost by drowning or by the loss of growing crops. Near the Bay of Bengal, tropical cyclones are likely to occur, causing considerable devastation. Thus the population of Bangladesh faces many natural difficulties whose solution would require the application of a high level of technology.

Mere population numbers, environmental problems and shortage of natural resources do not necessarily mean that a country is overpopulated. Unfortunately Bangladesh is a poor country with a very low level of technology.

About 90% of the population depend in some way upon agriculture and two-thirds of the country's area is used for farming. Rice is the main crop, grown on over 80% of the farm land at some time of the year. Bangladesh produces about 5% of the world's rice production, more in fact than Japan, but the yield per hectare is very low—less than one-third of Japan's. Jute is the leading cash crop. Bangladesh needs to import food grains. Yields could be improved considerably by the provision of a better water supply and fertilizers.

Farming techniques are generally primitive, most of the work being done by hand. Improvement is hindered by the very small size of the farms, which average about one hectare (about the size of two football pitches). This fragmentation of the farm land also hinders the setting up of co-operative schemes to provide irrigation water or to strengthen river banks in order to prevent flooding. Tanks (small reservoirs) for the storage of irrigation water have been allowed to fall into disuse. These small family farmers cannot afford to buy fertilizers or improved seeds, so crop yields are low.

Industry is mainly restricted to the primitive processing of agricultural raw materials from the farms (jute, rice, tea) or forest products. Such industries tend to be labour-intensive, using human labour rather than capital equipment. Hence they contribute little towards an increase in the productivity of labour.

Government family planning programmes have been set up mainly to extend the use of contraceptives, but also to try to raise the age of marriage in a country where it has been commonplace for marriage to take place at the age of about 14 and for women to bear six children in their lifetime. Much opposition has been encountered partly because children are needed to increase the family's work force and to replace other children who frequently die very young, and partly because the limitation of births is regarded as opposed to the teaching of the Islam religion.

GENERAL PROBLEMS OF UNDERDEVELOPED (DEVELOPING) COUNTRIES

The underdeveloped countries are usually referred to as the Third World.

Characteristics of developing countries

The chief characteristic of developing countries is their low standard of living. Figure 3.8 shows the countries that have the lowest living standards in the world (a GDP per capita of under $1000). These would be classified as 'developing' but some others, such as Mexico and Iran, whose GDP per capita rises just above $1000, would also be included.

Developing countries generally form a broad belt across the world stretching from South America, through Africa, to southern and south-eastern Asia. Most of them are located within the tropics. Many of them are much poorer than Figure 3.8 suggests, particularly in Africa where in a considerable number of cases the GDP per

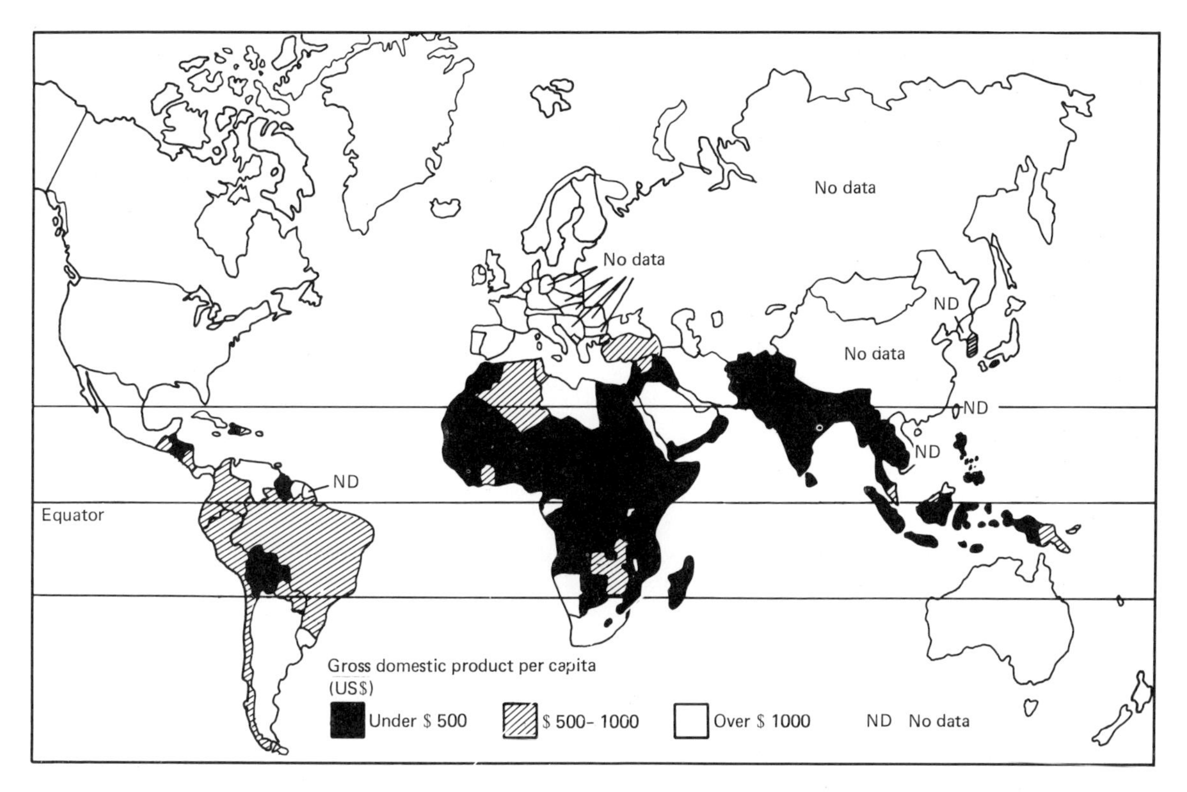

Fig. 3.8 Living standards in developing countries

capita falls below $100. There are also a few such cases in southern Asia (Afghanistan, Bangladesh), but living standards are generally rather higher in South America.

The great majority of developing countries have very high birth rates (Fig 1.7) and high (especially in Africa) or declining (in South America) death rates (Fig. 1.8). They are generally at stages 1 or 2 of the demographic transition (page 11) (Fig. 1.11), but a few, such as India, have progressed beyond this. A large proportion of their population is less than 20 years of age (often over half) and they have relatively few people over the age of 60 (see Fig. 1.14 for Bangladesh and Ecuador).

Many developing countries are still at stage 1 of the mobility transition (page 16) but a few, such as Pakistan, India, Bangladesh and some Latin American countries, seem to have reached stage 2. These tend to send emigrants to, for example, Great Britain and the USA.

Many developing countries, particularly in Africa, have mainly Negroid populations (pages 21 and 22. They are generally countries to which very few northern Europeans have emigrated in the last 500 years, but many southern Europeans have emigrated to Latin America.

Until the middle of the twentieth century most developing countries formed part of the empires of European countries, but they have now gained their independence and are often confronted by serious political problems (page 32). Many of them remain voluntary members of the Commonwealth (Fig. 2.6), including India, Bangladesh, Malaysia and many others in Africa.

The economies of developing countries are usually dominated by agriculture, which commonly employs over half of the labour force (Fig. 3.1) and well over 70% in some cases. Agriculture generally is not very efficient, and is usually carried on more as a way of life than as an economic enterprise. Mining is sometimes important, as in Bolivia, Iran, Iraq and Zambia for example. In some of these countries service industries employ a surprisingly large proportion of the labour force, but these are often rather trivial personal services which contribute little to the national economy.

Serious social problems often exist in developing countries as a result of the enormous difference in wealth and living standard between an affluent privileged class and the rest of the population. Hence it is not very satisfactory to measure living standards by means of an average value such as GDP per capita. Such contrasts in wealth often lead to serious internal political problems.

Many developing countries are regarded as overpopulated, even though population densities are often quite low. This is because technological progress has failed to keep pace with population growth in many cases.

In most developing countries economic progress is hampered by the lack of a social and economic infrastructure. There is often a shortage of railways, good quality roads and power supplies. Education facilities are often inadequate; up to 80% of the population may be illiterate. Hence it is difficult for people to learn new skills and become more productive. Such an infrastructure in advanced countries is created by the savings of the population or the taxes they pay. In developing countries people are mostly so poor that they are not able to save.

Possible solutions to the problems of developing countries

An increased supply of capital could help to improve agriculture, develop manufacturing industry and create an efficient infrastructure. In this way the technology developed in advanced countries could be transplanted to the developing countries. Help is already given to many developing countries by, for example, the USA, and recently China has begun to provide much aid to African countries. Assistance is given by a number of advanced countries to developing countries in south-east Asia through the Colombo Plan.

The advantage of importing capital from advanced countries is that it relieves the developing country of the need to sacrifice present living standards in order to create economic resources which will only help to raise living standards in the future. But some developing countries fear that, if capital is imported, they will lose control over their industries and may fall under the political influence of the country that supplies the capital.

A problem also exists in the way in which the capital is best used. To create capital equipment which simply saves labour is of little value, since most developing countries have a plentiful supply of labour. Capital should be used to develop natural resources which are still unused, thus increasing production. Projects such as the creation of irrigation systems, perhaps combined with the generation of water power, are of this type.

To help improve agriculture it has often been beneficial to set up a programme of land reform. This can release the farmers from exploitation by rich landlords and can permit the introduction of comprehensive schemes to improve the supply of water or to control soil erosion. It is possible to set up co-operatives controlled by the farmers themselves, to provide tools, fertilizers and to market the products. This could greatly improve agricultural productivity.

Industrial development could help to raise living standards, especially if the products were exported and the revenue used to buy imports and to create new capital assets. If capital for industrial development is difficult to obtain 'intermediate technology' can be used, i.e. the development of industries which use a large amount of labour and only comparatively simple and cheap equipment. Frequently, for example, the first industries to be created in a developing country are the manufacture of textiles and food products.

Pressure of population can be reduced by providing advice and aid in family planning, but such a policy may depend upon the development of education. Rural populations must at least be literate if a family planning programme is to be successful.

It has been suggested that restricting the growth of population is a better way of improving living standards than further economic development. It is argued that a reduction in the rate of population growth would allow resources to be transferred from present consumption to investment, so that more machinery and educational facilities could be provided. It could also help to reduce the great differences in personal wealth which exist in developing countries. Poor people tend to have the largest families, and their living standards could be raised if they had fewer children. It is also argued that such a policy would reduce the problems of pollution which often accompany economic development programmes, and it would also conserve natural resources for the use of future generations.

Development in a developing country—Ivory Coast

Ivory Coast, formerly a French possession, is situated on the Gulf of Guinea coast in West Africa. Its southern half has a wet equatorial climate and its northern half a savanna climate with a wet summer and a dry winter. It has few mineral deposits. Its capital and chief port is Abidjan (population 850000 in the mid-1970s) from which the country's only railway line runs north to Upper Volta (Fig. 3.9). Ivory Coast has enjoyed a remarkable success in economic development since the early 1960s. Its GDP per capita has risen from under $200 in 1961 to almost $400 in 1972 to over $600 in the mid-1970s.

Development has been initiated by the government rather than private enterprise. The government has tried to assess the costs and benefits of each project and has invested capital where the greatest benefits could be obtained for the country as a whole at the least cost. This has meant that most development has taken place in the Abidjan area where transport facilities are best and where there is the largest and most intelligent labour supply. Other parts of the country have tended to be neglected and people from these areas have tended to migrate to the Abidjan area. More recently some development has taken place along the railway line which leads north from Abidjan.

Little capital can be raised within the Ivory Coast because its people are generally poor, so grants and loans have been obtained from France, the World Bank and several European banks. Also capital has been obtained by earning revenue from exports, notably to France and other EEC countries. In general the government has tended to sacrifice principles of social equality in order to achieve the maximum possible economic growth.

Agriculture has been diversified from an excessive dependence on coffee and cocoa as export crops. In particular, the production of bananas, pineapples, palm oil and cotton has been developed, notably on the coast near Abidjan. The government has also encouraged the production of export crops by setting up government-controlled plantations and offering technical help, improved seeds and fertilizers to family producers near these plantations.

Greatest emphasis has been placed upon the development of industries for processing agricultural raw materials, such as the manufacture of food products, cotton ginning and the sawing of timber, but the textile industry has steadily developed. These industries can be developed with relatively little capital investment. More complex industries based on imports, such as chemicals, oil refining and steel manufacturing, have developed more slowly. Since the early 1960s there has been a distinct change in emphasis in the economy away from primary activities to secondary industries.

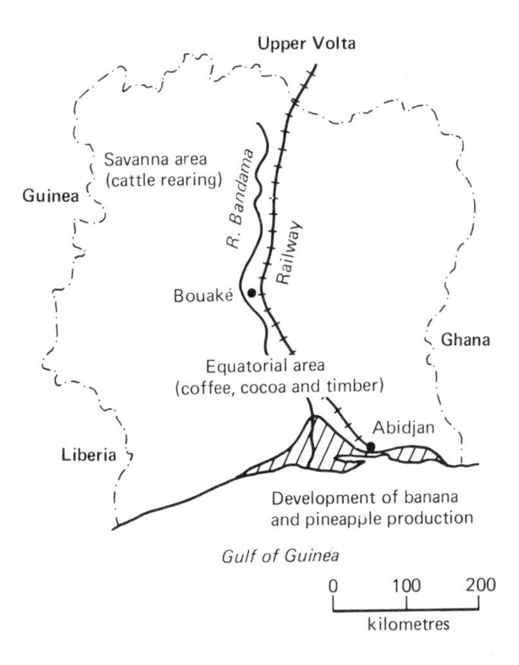

Fig. 3.9 Ivory Coast

Some problems have resulted from this development. Wages in manufacturing and service industries are very much higher than those in agriculture. Since these industries are mainly concentrated in Abidjan, many people have migrated from the rural areas to Abidjan. This has tended to drain young, educated people away from the rural areas and to cause unemployment in the Abidjan area. This problem has been increased by immigration from other African countries.

3.2 Regional problems of economic development

So far we have only considered the level of economic development of whole countries and the differences between them at the world scale. Although a country has an average level of economic development, shown for example by its GDP per capita (Fig. 3.8), there often exist great contrasts in development between different parts of a country. These contrasts give rise to what are called 'regional problems'.

CORE AND PERIPHERY

Economic development is hardly ever spread evenly over the whole of a country. It tends to be concentrated in certain areas which are known as core areas. Other parts of the country which are relatively undeveloped are known as the periphery.

Development would begin in the core with the introduction of some new economic activity, such as the establishment of a harbour or a number of industries. This causes a chain reaction which results in a steadily rising level of economic activity and growth. In the periphery, in contrast, little development takes place. Thus a great contrast develops between the busy, wealthy core and the stagnant, relatively poor periphery.

This process is illustrated in Figure 3.10. A new economic activity develops in one part of a country. This results in an increase in employment and probably of population in the area (arrow 1). Later, linked industries may develop (arrow 2). These may produce the raw materials for the new economic activity (backward linkage) or may use the new activity's products as its own raw materials (forward linkage). These linked industries also tend to increase employment and population in the area (arrow 3).

As employment increases so does the general level of wealth (arrow 4), partly through the development of these industrial linkages (arrow 5). This extra wealth may be used to expand the area's infrastructure (roads, railways, airports, buildings of all kinds, education facilities, etc.) (arrow 6) which becomes necessary because of the increase in employment and population (arrow 7). This, of course creates further employment opportunities (arrow 8) which further increase the wealth of the area (arrow 4).

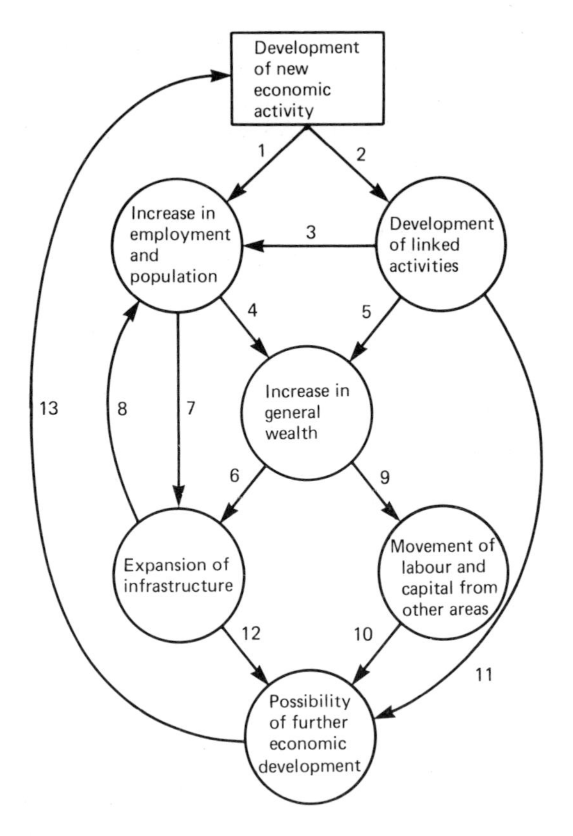

Fig. 3.10 Development of a core area

Besides financing the development of the infrastructure (arrow 6), the increasing wealth in this core area also attracts labour and capital from other parts of the country (arrow 9). The scene is therefore set for the further development of the core by the extra labour and capital that have moved in (arrow 10), the different industries that have already been established (arrow 11) and the improvements in the infrastructure (arrow 12). Thus new economic activities may develop in the core (arrow 13) and the whole process begins again.

It can be seen in Figure 3.10 that some of the arrows form continuous circular flows. Examples are arrows 4, 6 and 8, arrows 2, 11 and 13, and others. In these circular flows, an increase at any point spreads round the circle causing further increases at each point. They are sometimes called 'positive feedback loops'. They tend to cause steady uninterrupted growth throughout the whole system. The whole process of development illustrated in Figure 3.10 is sometimes called 'cumulative causation'.

THE EFFECT OF THE CORE'S DEVELOPMENT UPON THE PERIPHERY

As the core develops it is likely that economic resources, such as labour, capital and raw materials, will be drawn into it from the periphery, which is not developing to any great extent. This is known as a 'backwash effect'. It is also possible that some of the core's increasing wealth will 'spill over' into the periphery and help to cause economic development to begin there. This is known as a 'spread effect'.

Backwash effects

Manufacturing and service industries (pages 39 and 40) (high value-added industries) tend to concentrate in the core, but they often use raw materials which are produced in the periphery. Hence, the periphery's economic activity may be restricted to primary, extractive industries such as mining and agriculture. Income will flow from the core to the periphery to pay for these primary products, but it will return to the core as the periphery's population buys the much more expensive manufactures exported from the core. Thus, the periphery is left with little wealth for the development of an infrastructure. Hence, transport facilities and social services, such as health and education may be of poor quality. Also the availability of a great variety of work in the core may attract younger and better educated people from the periphery, leaving it with many older, unskilled people.

Backwash effects therefore tend to increase the economic contrast (regional imbalance) between the core and the periphery. This is commonly the case in the developing countries of the world.

Spread effects

Theoretically it is possible for the wealth of the core to spread into the periphery and stimulate economic development there. This seems to happen only rarely in developing countries.

Spread effects are much more likely to occur in advanced countries where governments commonly carry out policies that are designed to stimulate economically backward areas. These often include the provision of grants to improve the infrastructure and to encourage firms to extend their activities in the periphery.

Even in advanced countries however, spread effects are usually strongest in the parts of the periphery that lie nearest to the core or near to the larger towns. Remote, sparsely populated areas are frequently unaffected by governments' regional policies.

CORE AND PERIPHERAL AREAS IN SOUTH AMERICA

The overall pattern

Figure 3.11 shows some of the variations in economic development in South America. The core areas are indicated by large cities or clusters of cities mostly situated along the coast or on plateaux near the coast. Outside these core areas the peripheries are mostly concerned with primary activities such as agriculture or mining. Much of the agriculture is for subsistence only, but commercialization has taken place in areas such as the Argentine Pampa and eastern Brazil. Much of the north-centre and far south of the continent, particularly the Amazon lowlands and the central and southern Andes, is practically undeveloped and unpopulated.

National boundaries generally pass through these unpopulated areas. Thus, in most countries, one can recognize at least one urbanized core area near the coast, backed by a relatively undeveloped periphery. Road and rail networks are generally fan-shaped, converging on the core areas. Figure 3.12 represents a model of the spatial pattern of development in most South American countries. An outstanding example is Argentina, where nearly 40 % of the population, and an even greater proportion of the manufacturing and service industry, are concentrated near the Plate estuary. In Brazil, the main core area is situated in the south-east and includes the great cities of Sao Paulo and Rio de Janeiro, both with over seven million people in the early 1970s.

From Venezuela in the north, along the west coast to Chile and Argentina in the south, the core areas occupy locations selected by the Spaniards some 400 years ago. In Brazil they were established by the Portuguese (page 27). Cities such

Fig. 3.11 Variations in regional economic development in South America

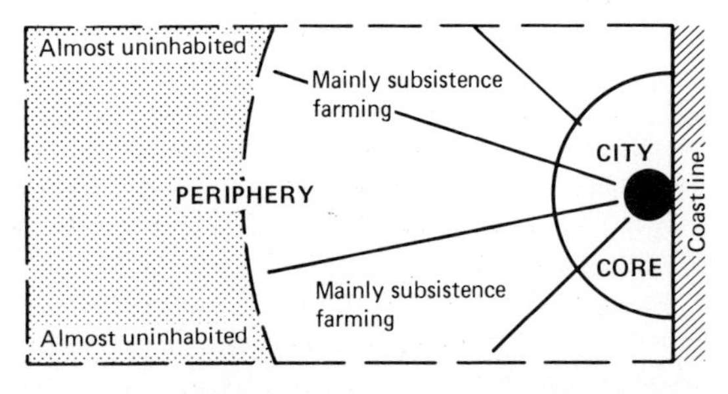

Fig. 3.12 Spatial pattern of development in a South American country

as Buenos Aires and Rio de Janeiro became bases from which the resources of the interior were exploited. Profits have been used mainly to enlarge and to develop infrastructures near these major cities. Backwash effects have been far more important than spread effects. Primary products have been obtained from the peripheries but most of the manufacturing industry and administration has remained in the cores. Governments in general have not seen the need to develop major projects outside the cores because there has been little pressure of population on resources.

Recently, however, a tendency has appeared to develop the periphery. Brazil, for example, has moved its capital city from Rio de Janeiro on the coast to Brasilia in the interior and has begun to develop a road network in the Amazon forests. The Brazilian government has actively encouraged investment in the development of its interior. Also, national governments have begun to feel that, in order to preserve their national territory they need to occupy it or develop it right up to the line of its frontier with neighbouring countries.

Backwash and spread effects in Venezuela

The development of the oil resources of Venezuela from the 1920s has helped the country to gain the highest living standards of any country in South America.

For many years this wealth was concentrated in the north of the country with a number of core areas of development extending from the major oilfield near Maracaibo (Fig. 3.13) to the capital Caracas. Much of the remainder of the country was populated by poor subsistence farmers and cattle herders especially in the Orinoco valley and in the Guiana Highlands to the south. Backwash effects led to the migration of population into the Maracaibo–Caracas belt.

Fig. 3.13 Venezuela

About 1960 the Venezuelan government began a project to generate economic development in the Orinoco and Caroni valleys (Fig. 3.13). The aim was to increase the total wealth of the country and also to reduce congestion in the northern core areas. The Guayana Corporation was set up to co-ordinate this development. The selected area possesses iron ore and oil and gas resources, and a new city, Santo Tomé de Guayana, was set up at the confluence of the Orinoco and Caroni rivers. This was supplied with power from the nearby natural gas fields and a hydro-electricity station on the Caroni river. Good roads were built linking the area to Caracas, and the Orinoco was dredged to permit it to take large steamships.

In addition, drainage and flood control schemes were begun in the Orinoco delta with the object of introducing cash crops such as sugar cane, bananas and cotton. Co-operatives were set up to supply farmers with seeds and fertilizers. Also, outlying areas in southern Venezuela near the boundaries with Brazil and Colombia have been provided with road transport and airstrips to integrate them with the remainder of the national territory. These events in Venezuela are a good example of the government encouragement of spread effects.

POPULATION DRIFT TO THE CITIES IN DEVELOPING COUNTRIES

Causes of the drift to the cities

As the core develops there is a tendency for population to move into it from the periphery. This is caused partly by forces which encourage people to move away from the countryside ('push' influences) and other forces which attract them to the cities ('pull' influences). This kind of population migration tends to occur during stages 2 and 3 of the mobility transition (page 17) when population is increasing rapidly.

The general poverty, hard manual work and monotony of rural life tend to encourage migration. Because of traditional systems of land ownership it may be impossible for a family to obtain more land and thus raise its living standard.

Accidents of the weather such as floods and drought may compel families to abandon their farms.

The modernization of farming through the introduction of mechanization or a change from subsistence to commercial farming may reduce the demand for farm workers. Thus many people become unemployed or underemployed. The mechanization of farming in the nineteenth century in England resulted in the expansion of towns as people left the rural areas.

Sometimes insecurity in rural areas may encourage people to move into the towns. Refugees flooded into Indian towns from Pakistan when partition took place in 1947. More recently, many rural inhabitants took refuge in Saigon (now Ho Chi Minh City) during the Vietnam War.

As education spreads into rural areas in developing countries it widens the experience of the people and causes them to want a fuller life. Hence, it is common for the better educated people to move into the towns.

As explained on page 48, in developing countries economic activity tends to become increasingly concentrated in the cities, and particularly in the larger cities. Urban wages are generally higher than rural ones. People migrate in the hope of improving their living standard by obtaining work in the growing industries.

Women in particular are attracted to the cities in search of work in domestic service for example. Wage-earning opportunities for women hardly exist at all in rural areas.

Information about the city's attractions has become much more likely to spread into rural areas. Transistor radios are almost universal and they tend to supply information about life in the city to rural communities. Also, bus services have expanded, and this has made it much more likely that residents of the countryside will visit the cities and take back accounts of their attractions to their home villages.

It is believed that the 'pull' of a city decreases steadily with distance. Hence the city is most likely to attract migrants from nearby rural areas. Further away from the city its attraction decreases, but also, people who live in large villages or small towns are more likely to migrate to the city than those who live in small villages and hamlets. This may be because education facilities are better in the larger rural settlements. Figure 3.14 illustrates this principle. The city attracts migrants from all sizes of rural settlements which are situated close to it. At greater distances only the larger rural settlements are likely to supply migrants. This is called a 'distance decay' principle.

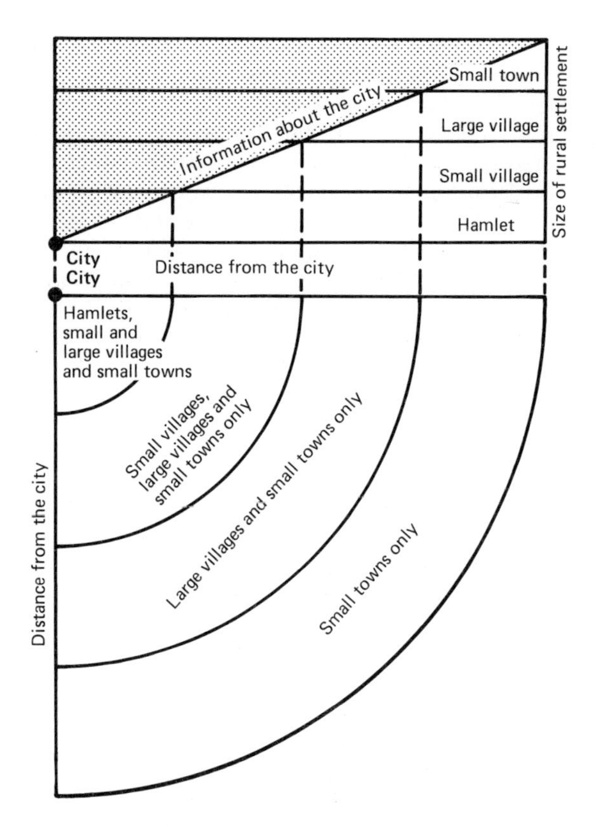

Fig. 3.14 Rural-urban migration in a developing country

Effects of the drift to the cities

In some areas, when people migrate to the cities, some relief may be given to the problem of overpopulation. Most emigrants tend to be of child-bearing age, so that the birth rate in rural areas may be reduced, so that population growth is slowed down. On the other hand, when rural areas lose young energetic people, and the average age of the population is thereby increased, there is a danger that the quality of agriculture may decline. In some cases it is the better educated and more intelligent members of the rural communities who migrate to the cities. Thus, villages may be deprived of their natural leaders and economic activity may stagnate.

City populations expand rapidly partly because of the large numbers of migrants moving in, and partly because these relatively young people tend to increase the birth rate and reduce the death rate of the city population as a whole. The populations of large cities have been known to double in under

10 years. This means that it becomes impossible for the city authorities to provide adequate housing or educational and health facilities. It may also prove difficult for immigrants to obtain regular employment, and they may be compelled to accept service occupations of very low status such as car washing, shoe shining and newspaper selling.

Living standards may decline disastrously in the city. For many years thousands of people have slept on pavements and railway sidings in Indian cities. Also slums develop especially near the city centres. These are crowded tenements consisting of old, decaying property, in which it is common for a family to occupy only a single room. In these slum areas vice and crime flourish.

The built-up area of the city may expand by the creation of shantytowns (known as 'barriadas', 'favelas', 'bidonvilles' and 'bustees' in various parts of the world). All major cities in developing countries have their shantytowns, which are ramshackle collections of buildings, usually occupying urban land which has not been used for any other purpose because it is too steeply sloping, subject to flooding or located on the outskirts of the town.

People move to the city from the countryside and, being extremely poor, may decide to build a rough single-room shack on an area of vacant land. If they find it difficult to obtain work they may cultivate small patches of land or rear a few livestock. Thus the shantytown may form a useful link between rural and urban life.

Although squatting of this kind is usually illegal, so many people may settle that the city authorities may hesitate to clear the area because of the problem of rehousing its inhabitants. Thus, over a period of time the shantytown dwellers enlarge their houses, build fences around their plots of land and create an irregular 'rural' type of settlement pattern in the city.

There are usually no industries in the shantytown, but shops, often grocers or bars, appear. Essential services, such as water supply and sewage disposal are often very primitive. Commonly a strong community spirit develops in the shantytown and its residents may elect committees to negotiate with the city authorities and thereby gain improvements in water, sewage and power facilities and even schools and permanent roads.

Shantytowns at an early stage of development are often a serious problem in developing countries because of the serious health and fire risks. Nevertheless many of them perform an important function in helping to adapt rural immigrants to urban life. Immigrants frequently live in the shantytown until they have accumulated some savings, when they will move into the city's normal residential areas.

REGIONAL PROBLEMS IN GREAT BRITAIN

Problems concerning the unevenness of economic development exist in advanced countries as well as in developing countries, but they are not always relatively simple contrasts between a well developed core area and an undeveloped periphery. Such problems exist in Great Britain and in other European countries.

The nature of the problems in Great Britain

Figure 3.15 shows the general pattern of the distribution of prosperity in Great Britain. Prosperity is generally greatest in a broad belt extending from the Midlands to south-east England generally along the routes of the M1 and M6 motorways and the electrified London–north-west England and Scotland railway route. With increasing distance from London the level of prosperity generally decreases westwards towards Cornwall and Wales, and northwards into northern England and Scotland. The least prosperous areas of all are the old industrial areas of central Scotland (Glasgow area), north-east England, Merseyside and the South Wales coalfield.

In the peripheral areas of the west and north, relatively low living standards are reflected in high unemployment rates. In times of recession the percentage of the labour force that is unemployed generally increases with distance from London. These areas also tend to have a lower standard of health services, higher death and infant mortality rates, and a greater proportion of inadequate houses than most of south-east England outside London. Also a smaller proportion of children tend to stay on at school and continue their studies at universities than in the south-east generally. Many of the towns of the north and Scotland in particular have serious problems of urban decay which are generally absent from rural areas of the south-east.

In the Midlands–south–east England belt incomes are generally higher and unemployment is usually less serious. Hence for many years people have tended to migrate from the west and the

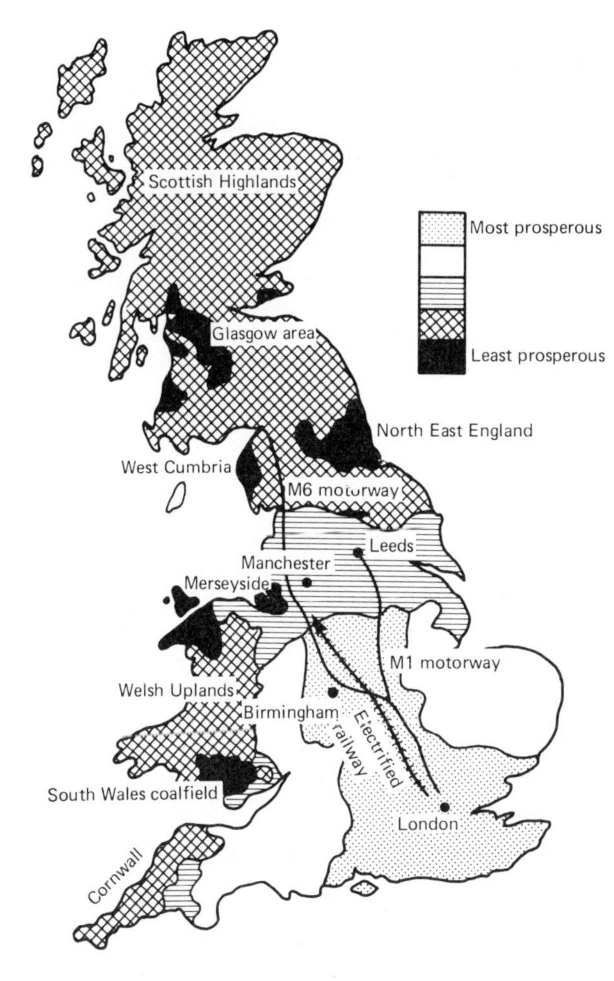

Fig. 3.15 Great Britain–variations in prosperity in the mid-1970s

north to the south-east (the 'drift to the south'). Despite this migration unemployment rates remain comparatively high in the west and north.

Reasons for the regional differences

During the nineteenth century a great development of manufacturing industries took place mostly on the coalfields of western and northern Britain. These included coal mining, iron and steel (in South Wales, Yorkshire, north-east England and the Clyde valley), textiles (in Lancashire and Yorkshire), and shipbuilding (in Merseyside, north-east England and Clydeside). These were based generally upon the availability of coal and other raw materials. Many towns came to be excessively dependent upon a single industry, such as the coal mining towns of South Wales, the cotton towns of Lancashire and the shipbuilding towns of Tyneside.

In the twentieth century manufacturing industry has begun to be attracted towards the markets for its products instead of towards its raw materials. The importance of coal as a power supply has greatly declined as oil and natural gas have come into use. Britain has suffered competition from overseas countries in the textile and shipbuilding industries. Thus the basic industries of west and north Britain have tended to decline and these areas have tended to suffer from unemployment.

For several centuries London has attracted industrial development. As the capital city and the largest concentration of population in Britain it has provided both a labour force and a market for products. In the twentieth century, industry developed in mainly rural areas in the zone between London and north-west England. This area became known as the Coffin Belt, because of its shape. It had a major port at each end, London and Liverpool, and enjoyed the best road and railway transport facilities in the country. Many new manufacturing industries, such as the motor car and electrical goods industries, grew up here, taking advantage of the excellent transport facilities. These industries attracted labour from western and northern Britain and the population of south-east England increased. This provided a growing market for other industries such as the manufacture of food products. In the twentieth century many different industries tend to locate themselves near to markets for their products. Population here has continued to grow and the building of motorways has increased the area's advantages of transport, so manufacturing industry has continued to develop.

In Britain and other advanced countries, during the twentieth century, there has been a great increase in employment in tertiary and quaternary occupations (page 39). Now well over half of Britain's employed population work in services of various kinds, often in offices and shops. In particular there has been a great increase in office employment in banking, insurance, trade and various government services. About half of the office workers of England and Wales live and work in south-east England. Even in times of economic difficulty there is usually little unemployment in office occupations. They also contribute towards the generally higher wages that are received in south-east England.

Government regional policies

Two different attitudes may be taken about regional differences in economic development. The government may feel that development should be allowed to continue in the prosperous region despite its effects on less favoured regions, since it will create wealth which will increase the country's overall prosperity. On the other hand, the government may feel that continued growth in the prosperous region will lead to overdevelopment and congestion and that social justice demands that economic activity should be channelled into the poorer regions. In Britain the government has generally taken this latter view. Its policies have included the following.

The government has provided incentives for employers to set up firms in the main problem areas. Various parts of west and north Britain have been classified as development areas or special areas. Grants, loans and tax concessions have been offered to industrialists setting up in these areas. Vehicle assembly plants have been set up, for example, in Central Scotland (e.g. Linwood, which has now closed) and Merseyside (e.g. Halewood) and new steel works have been established in the Clyde Valley (Ravenscraig) and South Wales (Llanwern) At the same time the government has tried to restrict expansion in the south-east by controlling the building of factories by the use of Industrial Development Certificates. The Location of Offices Bureau, created in the 1960s, used to encourage offices to leave central London.

Transport facilities have been improved in the peripheral areas by the creation of motorways, new harbours, Inter-City and Freightliner rail services and regional airports. These have tended to encourage the growth of modern industries.

New towns have been established and developed in problem areas such as central Scotland (e.g. Cumbernauld and Livingston), north-east England (e.g. Peterlee, Newton Aycliffe and Washington), north-west England (e.g. Skelmersdale, Runcorn, Warrington and the Central Lancashire New Town) and South Wales (e.g. Cwmbran). It was hoped that these towns would generate spread effects (page 49) and stimulate development in nearby areas. An attempt to relieve congestion in London has been made through the establishment of a ring of new towns around London (e.g. Basildon, Harlow, Welwyn Garden City, Stevenage, Hatfield and Hemel Hempstead to the north, Bracknell to the west, and Crawley to the south).

Exercises

1. (*a*) Describe the changes that have been observed to take place in a country's employment structure as it progresses to a higher level of economic development.
(*b*) To what extent can such changes be regarded as relevant to the economic problems of the United Kingdom in the late twentieth century?

2. If technological development tends to increase a country's optimum population density it seems reasonable to expect a positive correlation between national living standard and national population density. How far can such a relationship be identified at the present time?

3. The developing countries of the world are sometimes referred to as the 'South'. Discuss the extent to which national living standards are related to latitude.

4. In the light of the various models relating to population and economic development describe and explain the ways in which you would expect the present-day Third World countries to develop in the next 50 years.

5. Describe and explain the possible effects on regional economic development of the establishment of a tourist industry in an isolated, sparsely populated part of a developing country. Suggest locations where such a development has taken place.

6. In what ways do present-day regional problems in advanced countries differ from those in developing countries? Quote examples.

7. Figures 3.16–3.18 illustrate some of the social and economic characteristics of 14 widely differing countries.

(*a*) Describe and suggest explanations for the relationships illustrated in these three diagrams.

(*b*) Construct a graph for these 14 countries showing the relationship between GDP per capita and death rate. Explain the principles upon which you have based your answer.

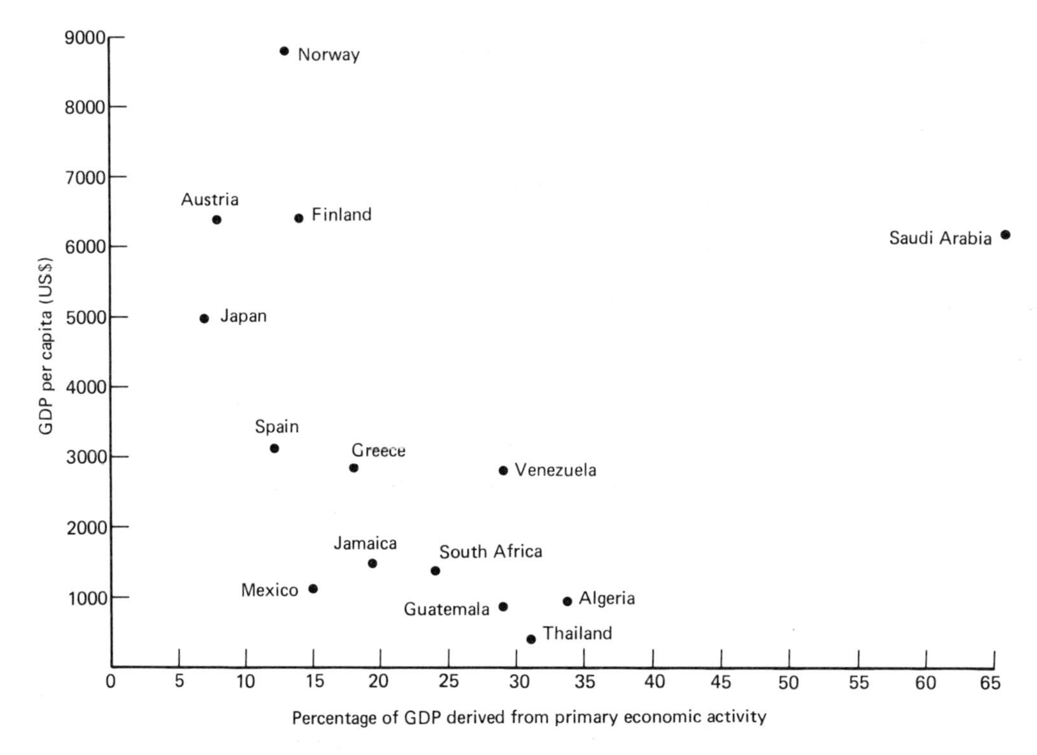

Fig. 3.16 Relationship between gross domestic product per capita and the importance of primary economic activity

(*Data derived from: UN Statistical Yearbook 1977, pp. 692–730. Copyright, United Nations (1978). UN Demographic Yearbook, Special Issue, 1979, pp. 171–175. Copyright, United Nations (1979). Reproduced by permission.*)

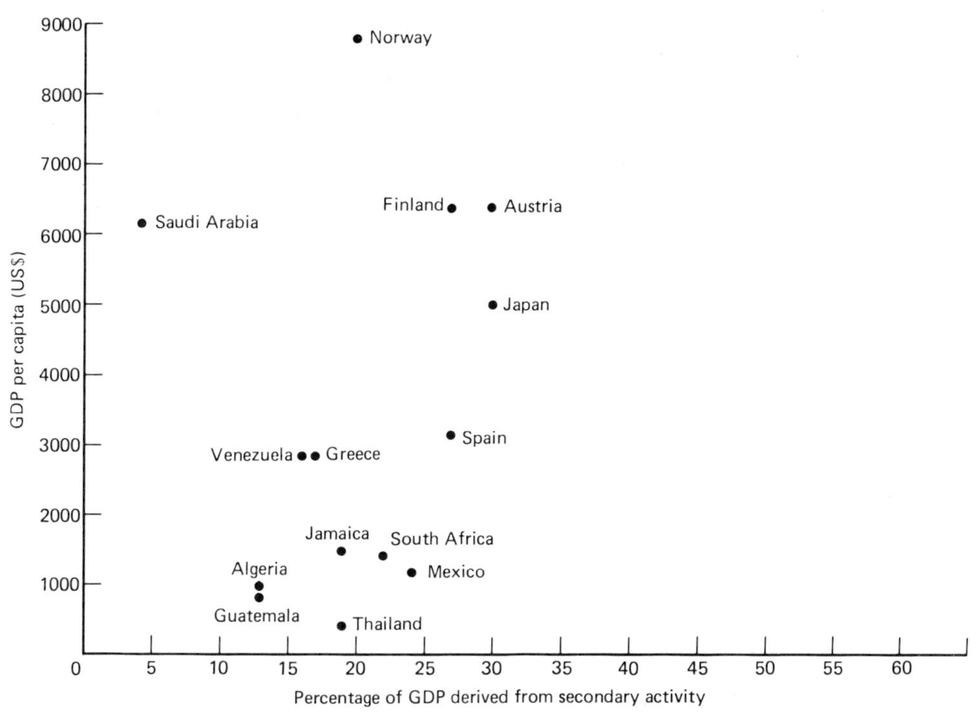

Fig. 3.17 Relationship between gross domestic product per capita and the importance of secondary economic activity

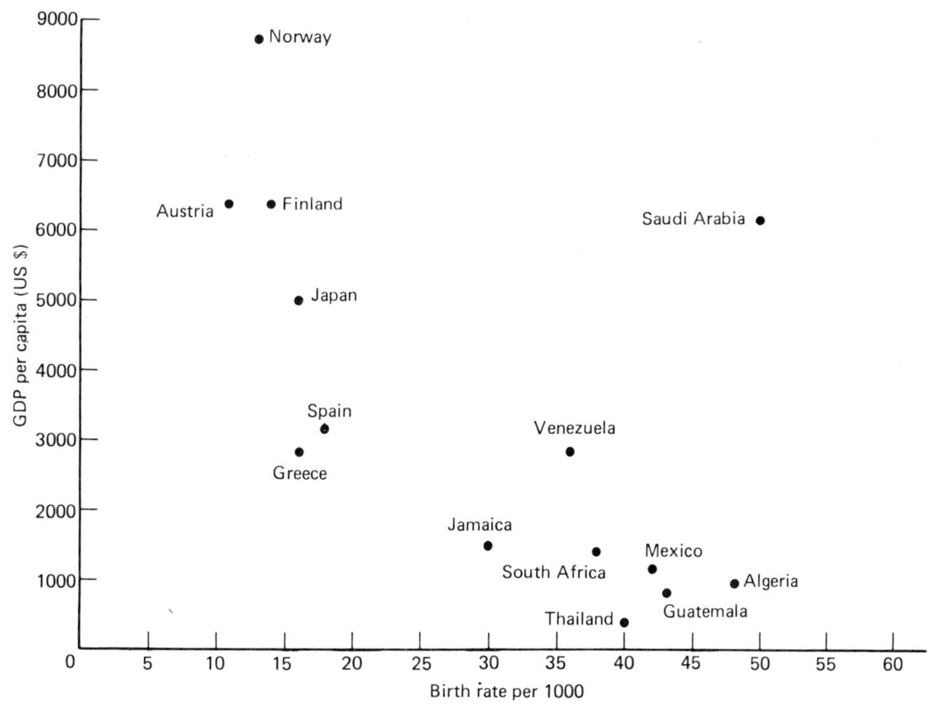

Fig. 3.18 Relationship between gross domestic product per capita and the birth rate

4 Farming

4.1 General principles

TYPES OF FARMING

Farming is one of man's 'primary' occupations (page 38). We have already seen how employment in farming has increased as the world's population has grown, but also that it employs a steadily decreasing *proportion* of the world's population. As the countries of the world have developed economically a greater and greater proportion of their populations comes to be employed in manufacturing and service industries.

Arable, pastoral and mixed farming
Farmers may concentrate on the production of crops (arable farming) or livestock (pastoral farming), but sometimes they may combine the two activities (mixed farming). In this case, crops are often grown to feed the livestock (Fig. 4.1(*a*)).

Livestock may be grazed on the existing natural grassland vegetation, but if crops are grown the natural vegetation is destroyed and an artificial vegetation (the crops) is grown in its place. This may cause the fertility of the soil to deteriorate, and it may be necessary to apply artificial fertilizers.

Since the mid-nineteenth century the area of arable land in the world has increased greatly, especially through the cultivation of the temperate grasslands of North America, Argentina, Australia and New Zealand by Europeans who have emigrated to these areas (page 11). This increase has now slowed down as fewer areas are left that can be made into arable land. Pastoral farming takes place mostly in those areas that are too dry, too cold or too hilly for successful arable farming. Many of these areas of overseas settlement have now been changed into areas of mixed farming.

Subsistence and commercial farming
Long ago, almost all farming must have been of the subsistence type (Fig. 4.1(*b*)), only providing for the needs of the farm household. The market for the farm produce was the farmers themselves. This was largely the case even in Europe only a few centuries ago in medieval times when the local woodland, pastures and open fields supplied most of the villagers' needs.

Gradually farming has become commercialized. Now in advanced countries, farmers do not consume all their produce but sell it to people who live elsewhere (Fig. 4.1(*b*)). The same farmers buy farm products from farmers who live in distant countries. Farmers in New Zealand, for example, produce butter to sell to people in the United Kingdom.

With the commercialization of farming there has also been an increase in *specialization*. In a subsistence system a wide range of crops and animal products had to be produced. Now many farms specialize in particular products. Dairy farms produce milk and cattle ranches produce beef, but subsistence farming still survives, especially in the less developed parts of the world along the equator and in south-east Asia.

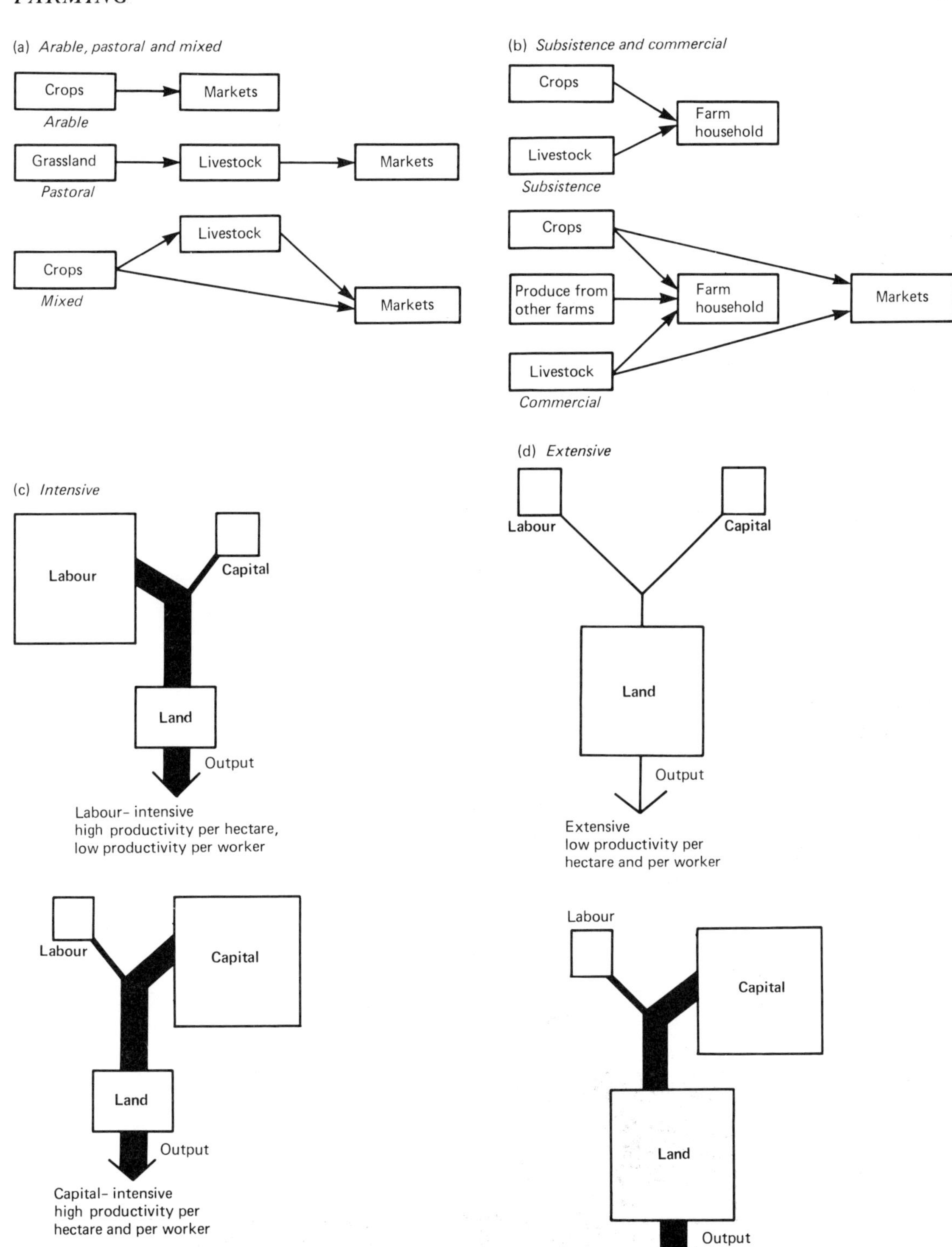

Fig. 4.1 Types of farming

Intensive and extensive farming

Farming involves the use of three different *factors of production*: land, labour and capital. Land is the land area of the farm together with its soils, and also its weather and climate. Labour is the work-force employed on the farm. Capital includes all the assets that have been created to make the farming more efficient, such as buildings, machines of various kinds, irrigation and drainage systems, improved breeds of plants and animals. Labour becomes much more productive if it is equipped with large amounts of capital.

Farming is said to be *intensive* when large amounts of either labour (labour-intensive) or capital (capital-intensive) are applied to a given area of land. In both of these cases productivity per hectare of land can be very high (Fig. 4.1(*c*)). Thus, a high productivity per hectare can be achieved by using either a great deal of labour or a great deal of capital. If much labour is used, productivity per worker may be very low, but if much capital is used, productivity per worker may be very high (Fig. 4.1(*c*)). Capital is to some extent a substitute for labour.

In *extensive* farming, labour and capital are spread rather thinly over the land. In primitive subsistence farming, for example, labour usually has the assistance of very little capital, so productivity both per hectare of farmland and per worker is usually very low (Fig. 4.1(*d*)). If on the other hand labour has the assistance of a great deal of capital, productivity per hectare of land may be low, but productivity per worker can be very high. An example of this is the commercial production of wheat in an area such as the Prairies of Canada.

A classification of types of farming

Farming types may therefore be classified according to whether they are:

(*a*) arable, pastoral or mixed;
(*b*) subsistence or commercial;
(*c*) intensive or extensive.

Such a classification is illustrated in Figure 4.2. In

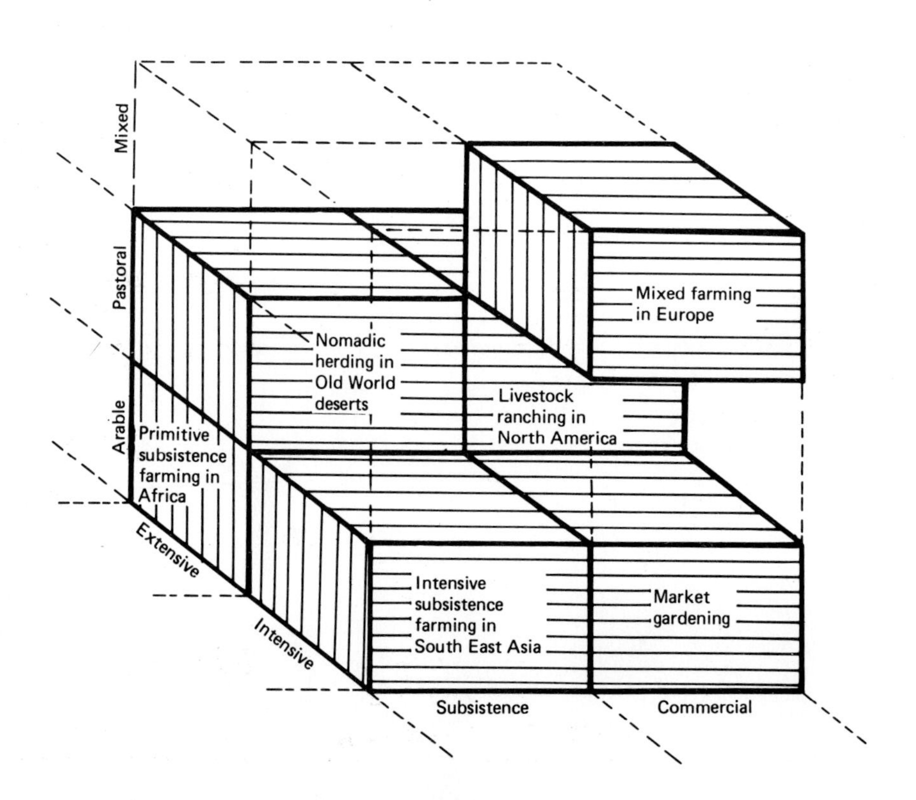

Fig. 4.2 A classification of farming types

this diagram, intensive subsistence farming in south-east Asia can be seen to be arable (the lowest level of blocks), subsistence (on the left-hand row of blocks) and intensive (on the front row of blocks). Both primitive subsistence farming and market gardening are also arable. Both market gardening and mixed farming are also intensive. Both primitive subsistence farming and nomadic herding are also of the subsistence type. The block which is hidden below the livestock ranching block could represent mechanized wheat production in Canada (arable, commercial and extensive).

THE INFLUENCE OF PHYSICAL FACTORS ON WORLD FARMING PATTERNS

Figure 4.3 shows the distribution of the world's main farming types. This distribution appears to have been influenced by the physical factors of climate and relief.

The influence of temperature

Very few crops useful to man will grow when the average temperature falls below 6°C, and many need much higher temperatures than this. Thus in Figure 4.4 it can be seen that in large areas of North America, Europe and Asia, there is a severe check to the growth of crops in winter when the average temperature falls below 6°C. In the tropics and in most of the southern hemisphere however, the temperatures permit the growth of crops throughout the year.

Crops also need a certain length of frost-free growing season in order to reach maturity. Very few useful crops are capable of reaching maturity if the frost-free season is less than 90 days. Thus most of Canada and part of northern Asia (Fig. 4.4) have too short a frost-free season for successful arable farming. These cold areas of the north also have the disadvantages of infertile podsolic soils and large areas where the subsoil is permanently frozen, a condition known as

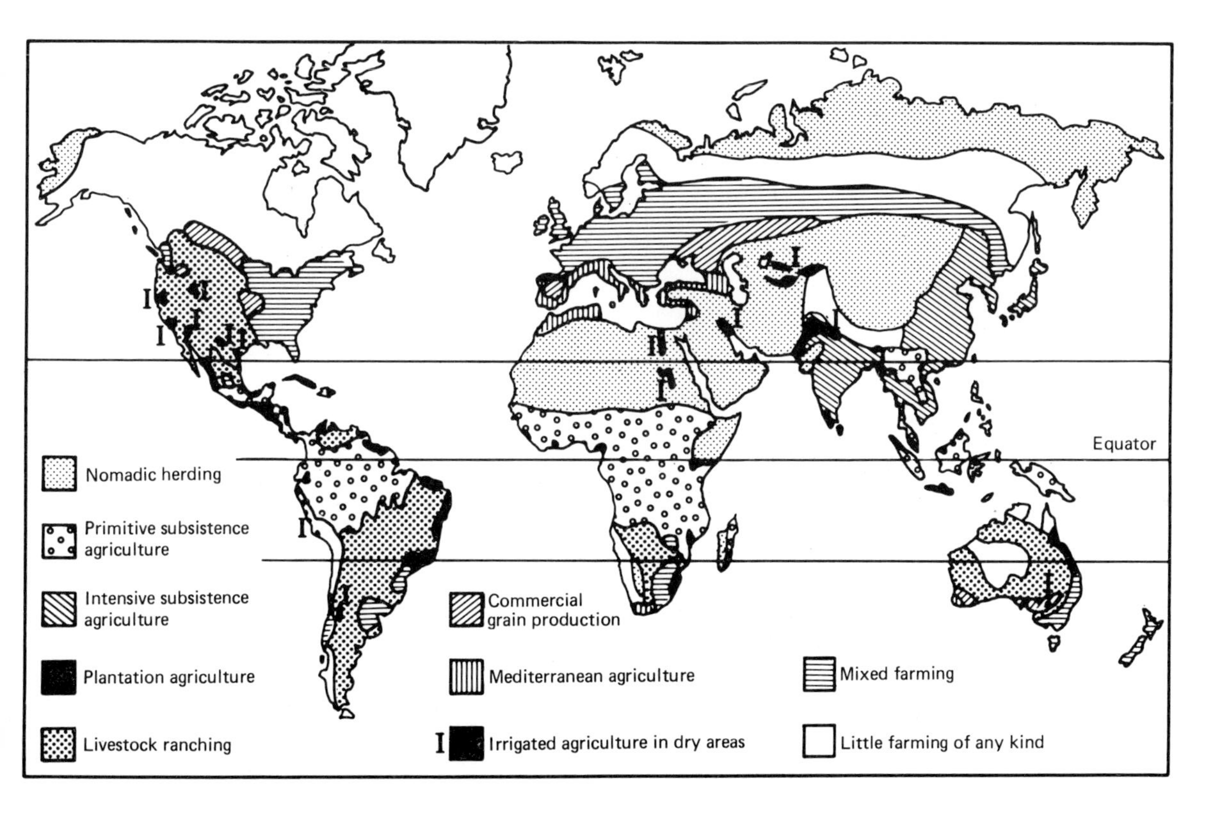

Fig. 4.3 Major world farming types

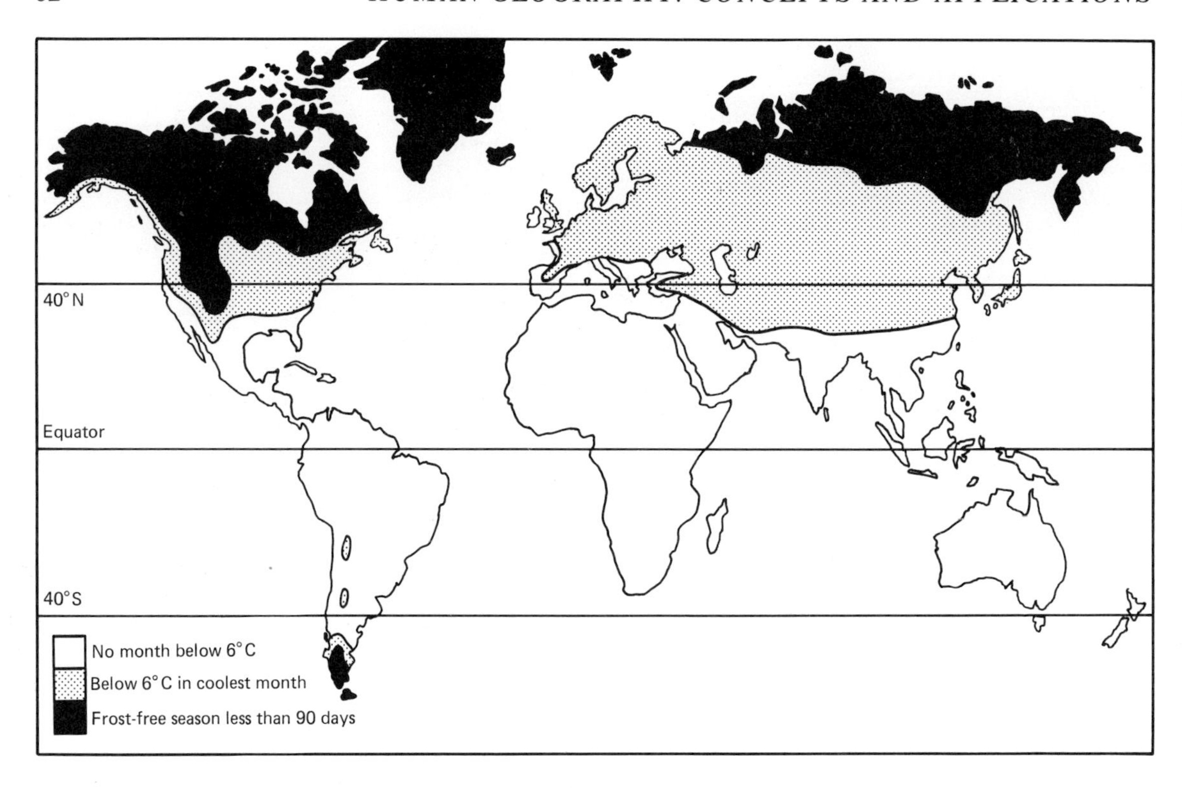

Fig. 4.4 Temperature in relation to farming

permafrost. To some extent the disadvantages of low temperatures and shortness of the growing season can be overcome by the breeding of quick-growing and quick-ripening varieties of plants. This has extended the area within which wheat can be grown in Canada for example.

The land areas of the southern hemisphere are much more favoured in respect of temperature than those of the northern hemisphere. This is partly because they are generally nearer the equator than the northern continents. In the southern hemisphere there is very little land to the south of latitude 40° south (Fig. 4.4). Also the large expanse of ocean in the southern hemisphere tends to moderate the climates of the relatively small land areas.

The influence of rainfall

Crops need an adequate amount of moisture, particularly during their growing season, but some crops ripen best if the weather is dry. Very few crops are able to grow if the annual rainfall falls below 250 mm, and in some parts of the world problems arise if the annual rainfall is less than 500 mm. Figure 4.5 shows the parts of the world where problems of drought are likely to influence farming. For a given rainfall total, these problems are more severe in areas where temperatures are high because much of the rain that falls is quickly evaporated. Hence 500 mm of rainfall might be reasonably adequate in Canada, but it would probably be inadequate in central Africa.

Crop growing is possible in very dry parts of the world provided that an alternative supply of water is available which can be used to irrigate the crops artificially. Figure 4.3 shows a number of areas in the west of the United States, on the west coast of South America, in the north-east of Africa and in the north-west of India, where crops are grown by irrigation in very dry areas. In all these areas rivers rise in nearly well-watered highlands (e.g. the Colorado, the Nile and the Indus) and flow across these desert areas where they provide water for farming. In semi-arid areas it is possible to use dry farming techniques. The land is cropped and left fallow in alternate years, so as to conserve water in

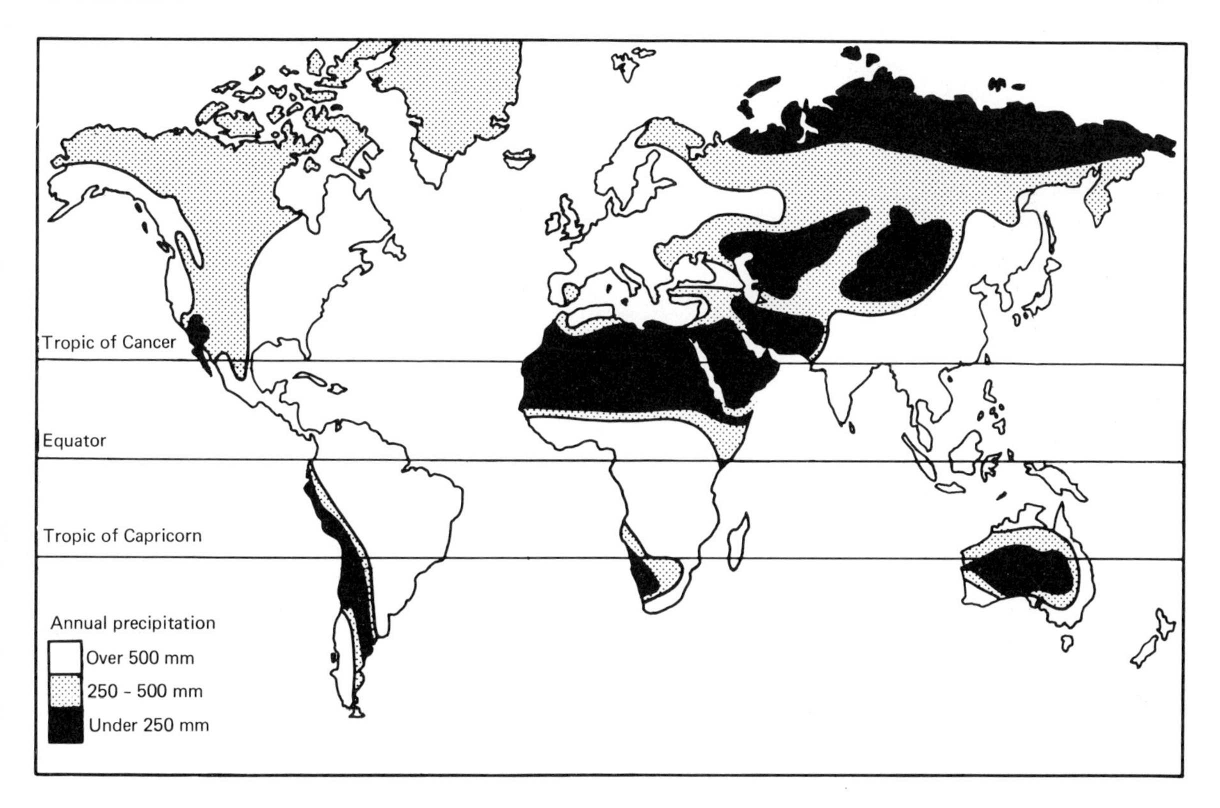

Fig. 4.5 Annual precipitation in relation to farming

the soil during the fallow year which may be used by the crop in the following year.

A serious problem of crop growing in dry areas is that, as the annual rainfall total decreases, its variability from year to year tends to increase. Thus it is possible to have a number of successive, unusually dry growing seasons which can be disastrous to farmers. Figure 4.5 shows that much of central Asia and northern Africa suffer severely from drought, but Europe is particularly well favoured. In the southern hemisphere Australia appears to have a greater problem of drought than either South America or southern Africa.

The influence of relief

Figure 4.6 shows the distribution of the main upland areas of the world. In these areas farming is likely to be hindered by the presence of steep slopes. Particularly in wet climates, heavy rain may erode the soil from even moderate slopes. Also, altitude has a very sharp influence on climate. With greater altitude average temperatures rapidly fall and the growing season quickly decreases in length, but plateau areas in the tropics may be more favourable to farming than nearby lowlands since both rainfall and temperatures may be reduced so as to suit a wide variety of crops.

There are two main belts of high fold mountains in the world. One runs parallel to the west coasts of North and South America; the other runs from south-west Europe, through the Middle East and the Himalayas, to south-east Asia. These fold mountains are very high and have large areas of steeply sloping land. On the other hand, the uplands of Africa are mostly relatively low plateaux.

The relations of farming types to physical factors

If temperature, rainfall and relief are considered (Figs 4.4–4.6) it appears that there are three particularly favourable areas for farming.

Lowland areas along or near the equator in South America, Africa and south-east Asia have high temperatures, a year-round growing season and rainfall totals over 500 mm. They are occupied largely by people engaged in primitive

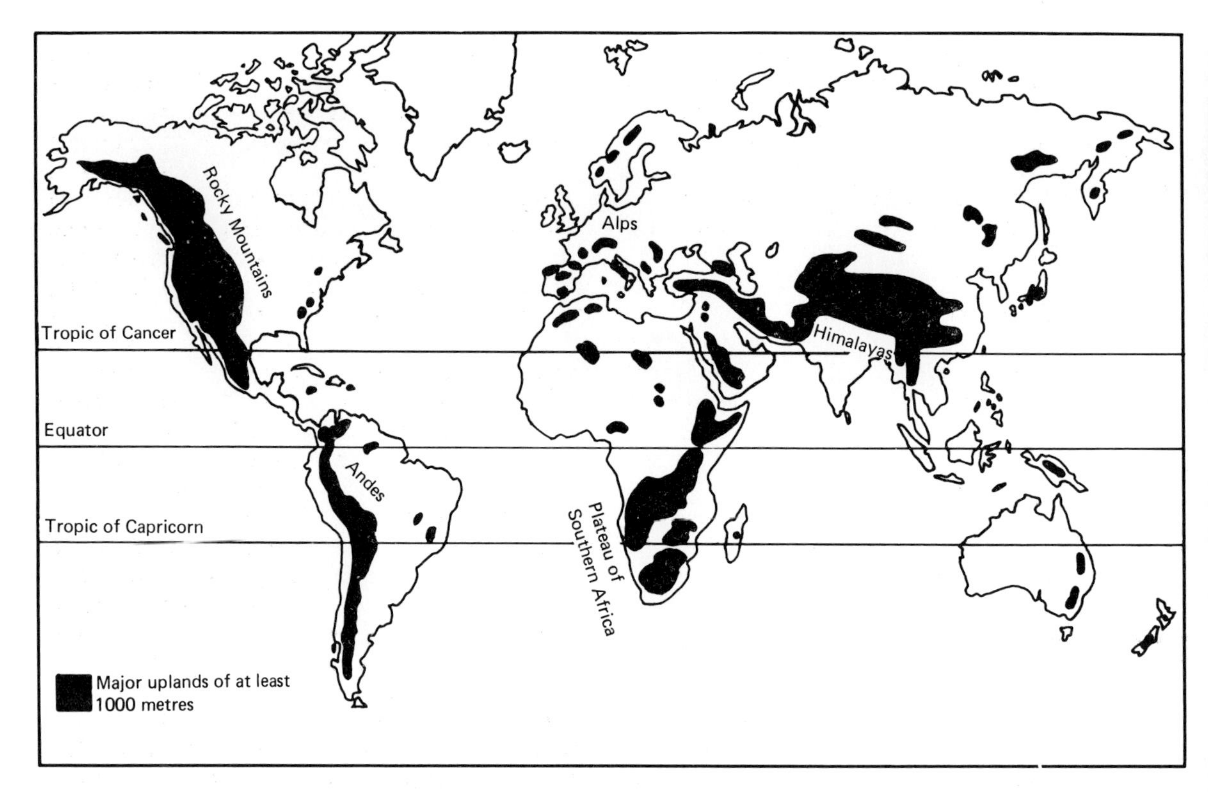

Fig. 4.6 Major relief features of the world

subsistence agriculture (South America and Africa) or intensive subsistence agriculture (south-east Asia). The real potential of these areas for commercial agriculture is not known.

The eastern parts of South America, South Africa and Australia also appear to have few serious disadvantages for farming. In these areas mixed farming has been developed by emigrant Europeans.

Eastern USA and much of Europe appear to have few disadvantages apart from comparatively low temperatures in winter. Here again, mixed farming is predominant.

In other parts of the world various problems exist, as in the following examples.

Western North America in general has cold winters, a short frost-free season and a relatively low rainfall. It also contains large upland areas. Extensive livestock ranching, with small patches of irrigation is characteristic here. Livestock ranching also predominates in the relatively dry parts of Australia, South Africa and South America.

Little farming takes place in the Old World Arid Belt between north-west Africa and central Asia. This area is dry and, in central Asia, mountainous and cold. Outside the patches of irrigated land the main type of farming is nomadic herding, though this is declining.

Farming, apart from a little nomadic herding, is largely excluded from the very cold northern coastlands of the northern continents.

THE INFLUENCE OF HUMAN FACTORS ON WORLD FARMING PATTERNS

The influence of urban markets and transport costs

Much of the world's farm produce is transported from farming areas to be consumed by people who live in urban areas. Hence farming is bound to be influenced by the location of these urban markets and the transport facilities that are available. A well-known model concerned with the influence of urban markets and transport costs upon farming was developed over a hundred years ago by von Thünen who owned and managed an estate in what is now East Germany. This section is based upon this model.

Von Thünen assumed the existence of an area of land (an 'isolated state') in which one central city was the only market for farm produce. This land area was assumed to be perfectly uniform in respect of relief, climate, soil fertility, etc., so that production costs per hectare for any particular crop were everywhere the same. Also, the transport costs for the crop increased steadily in proportion to distance in any direction from the urban market. The farmers of this uniform plain sent their produce to the central city and all received the same price for one hectare's production of any crop. These farmers all aimed to make the maximum possible profit, but it is clear that farmers near the city would make bigger profits than those further away because their transport costs would be smaller.

Let us consider first the distribution of a single type of farming (wheat production) on this uniform plain. Table 4.1 shows the price of a hectare's production of wheat when it is sold at the market (£160). The production cost per hectare is everywhere £40. The transport cost of a hectare's production of wheat is £20 for every 10 kilometres. Hence the total cost of growing and transporting a hectare's production of wheat increases by £20 for every 10 kilometres of distance from the market. Net profit is the difference between the market price and the total costs (production costs plus transport costs) per hectare. In this case, the net profit per hectare decreases by £20 for every 10 kilometres of distance from the urban market. The table shows that wheat can only be produced profitably up to a distance of 60 kilometres from the urban market.

Figure 4.7(*a*) shows all the information given in Table 4.1, and it is clear that net profit (economic rent) decreases to zero at a distance of 60 kilometres from the market, as a result of the rising transport costs. In Figure 4.7(*b*) only the net profit is shown. Figure 4.7(*c*) is a map of the area within which wheat can be grown profitably. It shows how the profit from wheat production decreases steadily outwards from the urban market to a distance of 60 kilometres. The extensive margin of wheat cultivation is the distance at which wheat cultivation yields no profit.

If changes occur in the market price or the production costs or the transport costs of wheat, profit levels may change and the area within which wheat can be grown profitably may either increase or decrease. Figures 4.8(*a*) and 4.8(*b*) show how profits may increase and the extensive margin of wheat cultivation may expand if either the market price rises or production costs fall. If the market price falls or production costs rise profits are decreased and the extensive margin contracts.

Figure 4.8(*c*) shows the influence of transport costs. These have little effect on farms located near the market, but a much greater effect on farms near the extensive margin. A decrease in transport costs allows more distant farms to grow wheat profitably; an increase in transport costs means that farms near the extensive margin become unprofitable. As an example, improvements in

Distance (km) from urban market	Transport cost (£) (per hectare)	Total cost (£) (per hectare)	Net profit (£) (per hectare)
0	0	40	120
10	20	60	100
20	40	80	80
30	60	100	60
40	80	120	40
50	100	140	20
60	120	160	0
70	140	180	–20
80	160	200	–40

Table 4.1 Wheat Production
Market price (per hectare) = £160
Production cost (per hectare) = £40

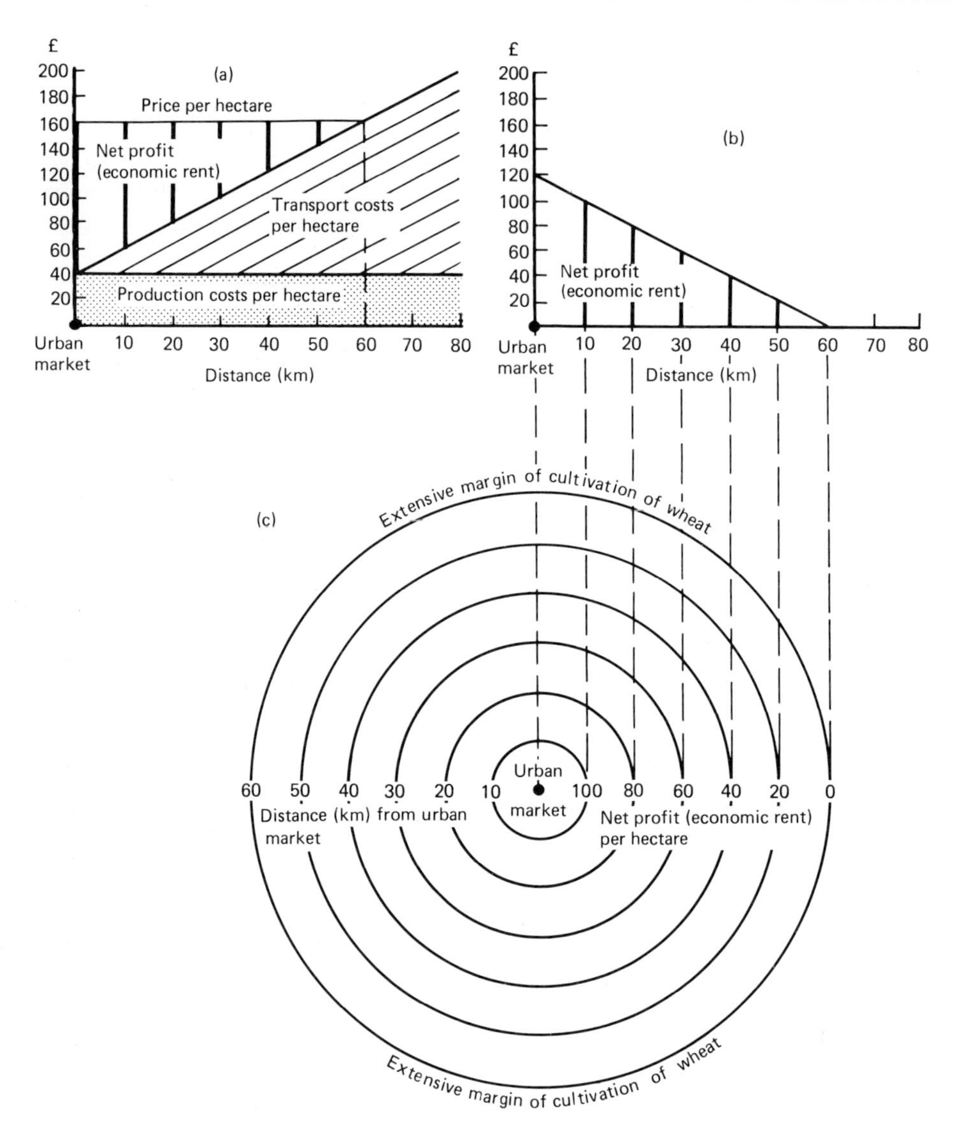

Fig. 4.7 Production of wheat

ocean transport in the nineteenth century allowed Canadian wheat to be marketed profitably in Europe.

We now introduce the possibility of other types of farm production in addition to wheat cultivation. Figures 4.9(*a*) and 4.9(*b*) show the price, the costs and the net profit earned from producing milk and wool. With milk, transport costs increase very rapidly with distance, so the net profit from milk production decreases rapidly to give an extensive margin at a distance of 40 kilometres from the market. Transport costs are much lower for wool production, so wool can be produced profitably up to 80 kilometres from the market.

If it is possible to produce either wheat, milk or wool in the area we have already considered, farmers will tend to concentrate on the type of farming that yields them the highest net profit. To discover this we draw net profit curves for wheat, milk and wool all on the same diagram (Fig. 4.10). This diagram shows that near the market it is most

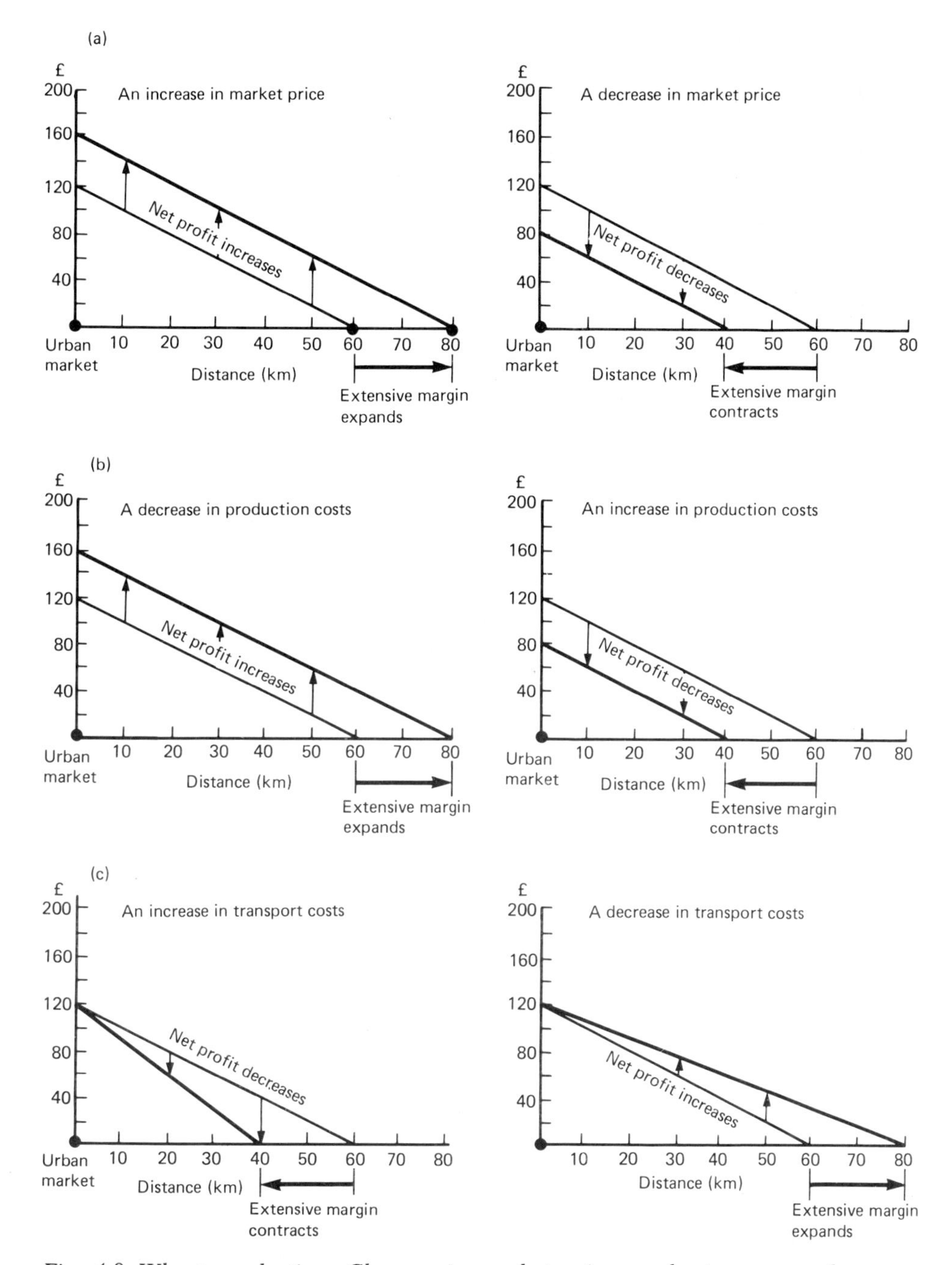

Fig. 4.8 Wheat production. Changes in market price, production costs and transport costs

profitable to produce milk, but at a distance of 20 kilometres from the market the net profit curve for wheat crosses that for milk, showing that wheat production has become more profitable than milk production. Farmers therefore specialize in wheat. At a distance of 40 kilometres from the urban market, wool production becomes more profitable than wheat production, so the farmers whose farms are located between 40 and 80 kilometres from the market specialize in wool. The points at

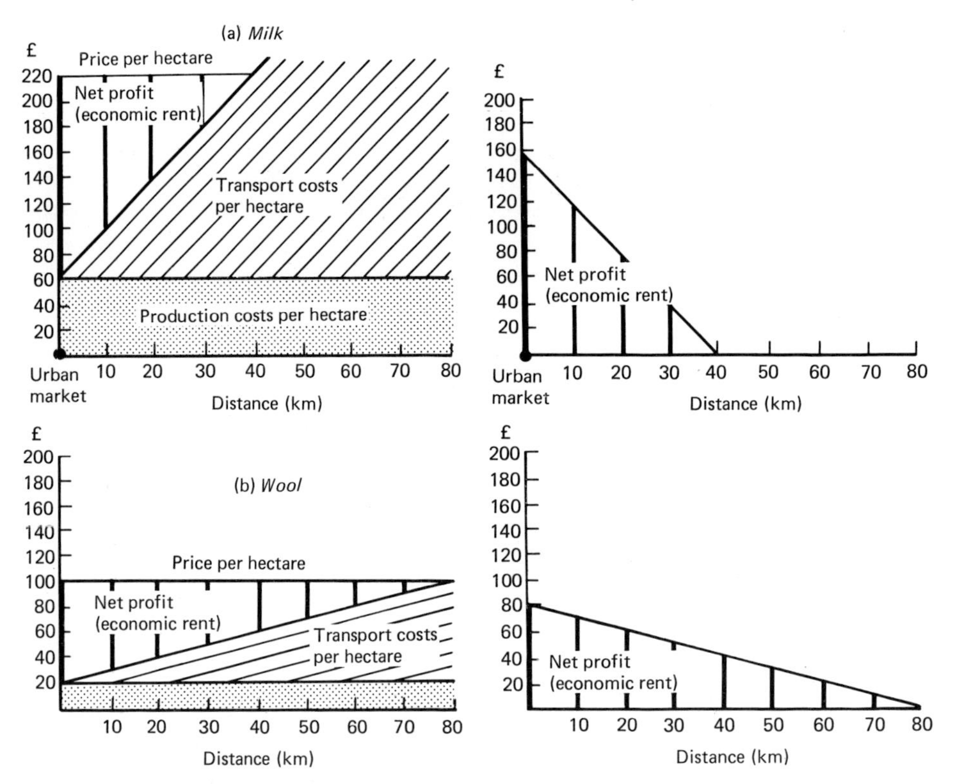

Fig. 4.9 Production of milk and wool

which one type of farming is replaced by a different type are known as *margins of transference*.

Figure 4.10 also shows a map of the types of farming within a distance of 80 kilometres from the urban market. It can be seen that milk, wheat and wool producing areas form concentric circles around the urban market. Profits are highest in the milk producing area and they decrease to zero at the extensive margin of wool production, 80 kilometres from the urban market. Because of these changes in the profit levels of the different types of farming it is likely that the *value of land* will be highest near the urban market and also that the *intensiveness of farming* will be higher nearer to the urban market. In the wool producing area, in particular, land will tend to be cheap and farming will be extensive.

In von Thünen's original model the sequence of rings consisted of milk and vegetables (perishable) nearest the central city, followed by timber (difficult to transport) and arable crops, with livestock forming the outermost ring.

If changes now take place in the market prices or the transport costs of milk, wheat or wool, these will be reflected in changes in the land areas devoted to these types of production. Since it is assumed that farmers aim to maximise their profits, they will change their type of specialization to that which yields the highest net profit. Figure 4.11 shows the effect of a decrease in the transport costs of milk. It can be seen that farmers in the inner part of the wheat belt have decided to change to milk production which, as shown on the graph, yields a higher net profit.

In fact of course conditions in the real world are unlikely to match this model exactly. Farmers for example are unlikely always to specialize in the form of production which at any given time maximizes their profits. This might mean that they were constantly changing their methods. However, the model is useful in indicating the general economic factors that have an influence upon farming. It is common for large cities to be surrounded by a zone of land in which dairy farming is of particular importance, and it is also common to find sheep rearing on land remote from urban markets, but consideration must also be given to the influence of physical factors upon agriculture (page 61), which this model completely ignores.

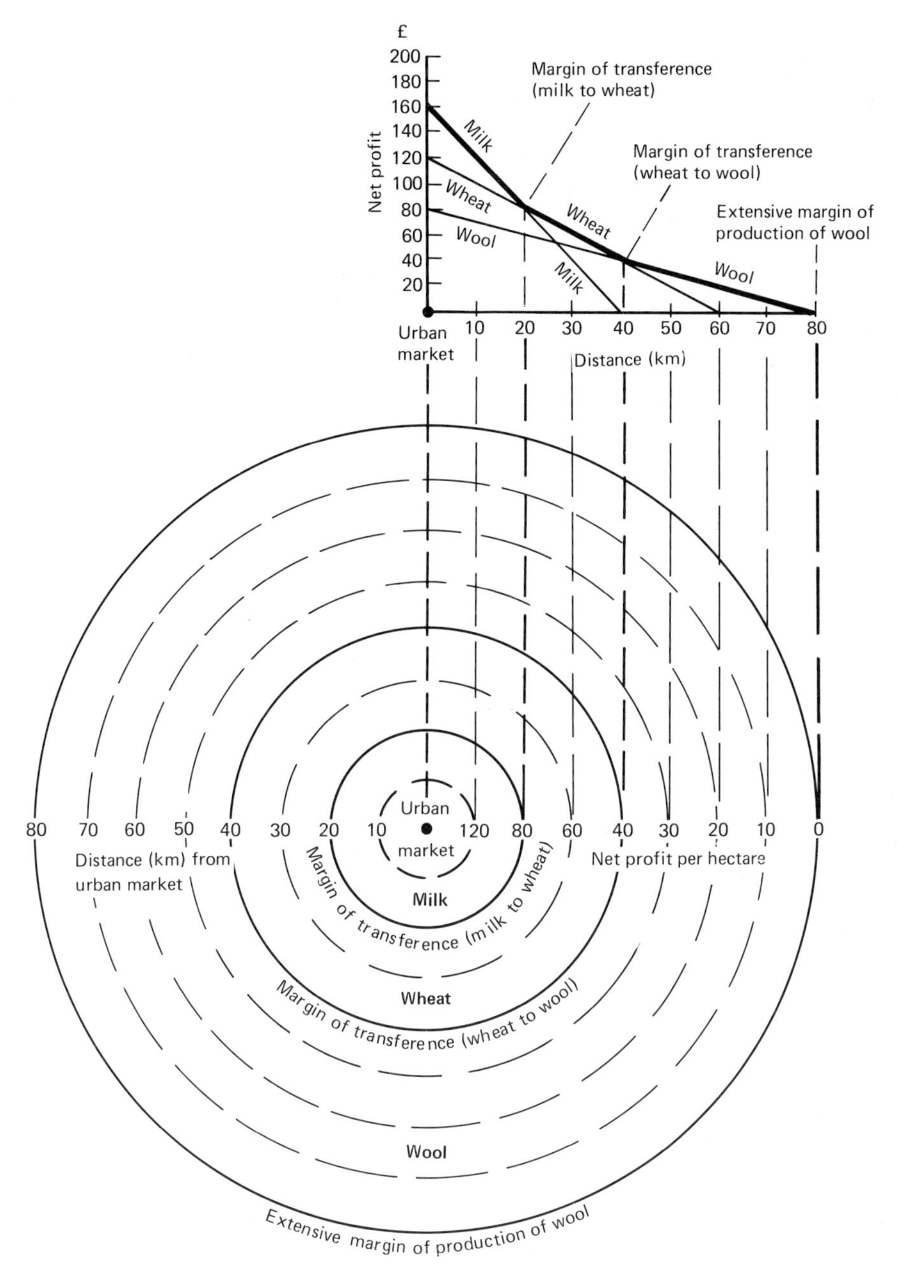

Fig. 4.10 Production of wheat, milk and wool

The influence of other human factors

Governments may interfere with the free-enterprise system described above in which farmers choose their specializations in the light of market prices and production and transport costs. Governments may increase the prices actually received by farmers by subsidizing certain farm products. This means that the farmer receives the price for which he sells the product plus an additional payment from the government. Alternatively, governments may restrict the import of certain farm products and thus force the

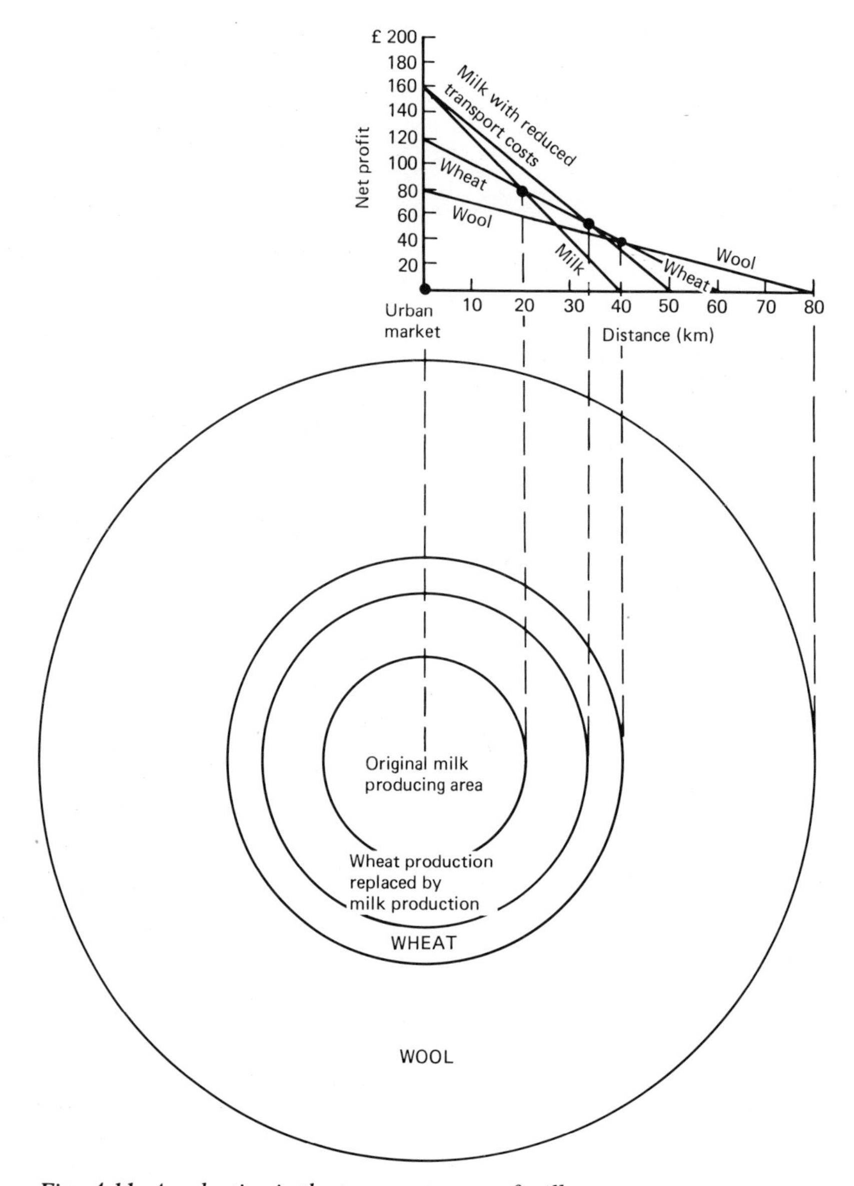

Fig. 4.11 A reduction in the transport costs of milk

price up sufficiently to encourage farmers to take up these types of production. Also, farmers' costs of production may be reduced by the provision of grants, loans and tax exemptions in order to persuade farmers to develop the production of certain crops.

Not all farm products are delivered by the farmer directly to urban markets. In some cases, the 'market' from the farmer's point of view may be a factory which either reduces the bulk of the farm product or changes it into a less perishable form. Thus, excessively bulky or perishable farm products tend to be produced near this processing plant, even if the plant itself is far distant from urban markets. Examples are sugar beet grown near processing plants, vegetables grown near quick-freeze plants, fruit grown near canneries.

The introduction of refrigerated shipping in the late nineteenth century stimulated the production of dairy produce in New Zealand and beef in

Argentina by making it possible for perishable products to be transported great distances. In Britain the Milk Marketing Board has stimulated milk production in areas such as Cornwall and south-west Scotland which are remote from sizeable urban markets.

The advanced countries of the world have comparatively large capital resources and so have been able to develop capital-intensive forms of farming such as dairy farming, market gardening, fruit production and mechanized arable farming. Also, great multi-purpose schemes have been developed to provide both irrigation water and electric power. In the USA comprehensive development projects such as those of the Tennessee and Missouri river basins have been embarked upon. Developing countries on the other hand lack capital. They have therefore had to substitute labour instead. This means that production per worker is much lower than in advanced countries and farming is largely subsistence in type.

Different types of farming vary greatly in their labour requirements. Orchards, market gardens and the production of greenhouse crops need large amounts of labour, at least at certain times of the year such as the harvest season. They tend to be located in densely populated areas, though in some cases it is possible to hire migrant labourers. The rearing of sheep and beef cattle, and mechanized grain production need little labour and can therefore succeed in very sparsely populated areas.

Farmers may be either owners or tenants in respect of the land that they work. If they actually own the land (freehold tenure) they may be encouraged to develop it and make it more productive. If they are tenants they may pay a money rent or a share of their crop (share-cropping) to the landowner. This may encourage them to work the land too intensively, with little thought for the future, and thus cause a decline in fertility. In the USSR state ownership and planning has created large state and collective farms which it was hoped would be extremely efficient as a result of their enormous size (economies of scale).

4.2 Farming in developing countries

GENERAL CHARACTERISTICS

In developing countries farming is very important and often employs over half of the labour force (page 38). It is mostly of the subsistence type (Fig. 4.1(*b*)) and each farm community is largely self-sufficient.

There is little specialization in farming. Each area and each farm tend to produce a wide variety of products most of which are usually foods. This is necessary because there is little trade.

Production depends to a great extent upon human labour assisted sometimes by that of animals. Capital is very scarce, so productivity per worker is usually low; but in areas of intensive farming productivity per hectare may be high.

Nomadic herding exists mostly in the arid belt which extends from north-west Africa to central Asia; primitive subsistence agriculture in the hot, wet equatorial regions of South America, central Africa and the East Indies; and intensive subsistence agriculture in eastern and south-east Asia (Fig. 4.3). In these areas Europeans have made comparatively little impact upon farming, apart from the development of patches of plantation agriculture and the development of irrigation schemes such as those of the Nile and Indus basins.

NOMADIC HERDING

Distribution

Nomadic herding is almost confined to north-east Africa and south-west Asia where the first animals were domesticated about 8000 years ago, but reindeer herders exist in the tundras of Europe and Asia. (Fig. 4.3). When Europeans first visited the Americas and Australia they met no animal herders, so it appears that the idea of domesticating animals never spread beyond Asia, Africa and Europe.

Characteristics

In the Old World arid belt the herders move between grazing grounds according to the season of the year. Their animals provide milk, cheese, butter, meat and hides, as well as transport.

Usually there is no permanent village; the whole community migrates with its herds, carrying tents and other possessions. There is no private ownership of land. The social group identifies itself with its leading members rather than with a particular area of land.

There are no clear limits to the area of migration. In poor years the group may range over a wider area than usual. Large areas of land are needed to provide sufficient grazing for the herds, so population density must be very low.

The type of migration is often influenced by the seasonal variations in climate. In the Zagros mountains of western Iran most of the rain falls in winter. At this season the nomads are living in the foothills. As temperatures rise in spring and the weather becomes drier, they move to higher ground following the snowline as it retreats. Here the melting snow and the lower temperatures help the growth of pasture in summer. As the winter rains begin again the nomads return to the foothills and plains to the south of the main Zagros range. A vertical movement such as this is sometimes referred to as 'transhumance'.

Nomads are usually an embarrassment to national governments because they have no settled homes. Hence many attempts have been made to settle them in specially built villages—a process known as 'sedentarization'.

PRIMITIVE SUBSISTENCE AGRICULTURE

Primitive subsistence agriculture (Fig. 4.3) includes *shifting agriculture* in which the cultivated plots are abandoned every few years and new plots created, and *sedentary agriculture* in which the same plots are cultivated year after year.

Shifting agriculture

Shifting agriculture was once common in Europe and was once practised by the Indians of North America. It now exists mostly in tropical forests and grasslands. It is known as 'milpa' in central America, as 'chitimene' in central Africa, and as 'ladang' in south-east Asia.

During a drier season of the year an area of forest is cleared by the use of axes. The felled trees are then burned and crops are planted in the mixture of wood ash and soil. Often the stumps of trees remain, and sometimes large trees are left untouched.

A great variety of crops is then planted using digging sticks or hoes. Basic food crops include maize, cassava, yams and bananas. Upland (unirrigated) rice is important in south-east Asia. The great variety of crops provides a vegetation cover which is not unlike the original forest.

These crops gradually use up the soil nutrients which had been built up by the original forest vegetation, so crop yields decrease rapidly, by as much as 50 % in the second year of cultivation. After three to five years the plot is abandoned and allowed to revert to forest. This particular piece of land cannot be cultivated again for at least 10 years.

Because of the constant need to clear new areas of forest each village needs a very large area of land. The village settlement itself has to be moved from time to time so as to be reasonably near to new clearings. Hence the maximum possible population density is about 10 per square kilometre, except in south-east Asia where upland rice can support a rather higher density.

Sedentary subsistence agriculture

Sedentary subsistence agriculture is found in the same general areas as shifting agriculture. In some areas shifting agriculture has evolved into this type of farming.

In true shifting agriculture both village settlements and cultivated plots are abandoned every few years and new ones are created elsewhere. In some cases, however, the settlements are permanent and only the cultivated plots are abandoned and re-created (a method known as 'land rotation'). In sedentary agriculture both the settlements and the cultivated plots are permanent. It is therefore an intensification of shifting agriculture.

The original vegetation is permanently destroyed and replaced by cultivated plants. Soil nutrients are used up as crops are harvested, so they have to be replaced by the use of manure and, if possible, artificial fertilizers.

This type of farming supports populations of moderate density especially on the plateaux of central Africa with sorghum (Guinea corn) as the main food crop, in the Andes of South America based on maize, and in upland areas of south-east Asia based on upland, unirrigated rice.

INTENSIVE SUBSISTENCE AGRICULTURE

Intensive subsistence agriculture (Fig. 4.3) has been a characteristic for many centuries of the valleys and coastal lowlands of the Far East from Japan to India. Traditionally the staple food crop has been rice.

Traditional rice cultivation in the Far East

In the Far East farming techniques have developed over a long period of time, and experience has been handed down from generation to generation. Rice is grown wherever possible because it is nutritious and stores well in hot climates. On level flood plains and deltas bunds are built to create paddy fields which can be flooded for the rice crop. The rice plant obtains some of its nutrients from irrigation water. Also the flooded rice fields can be stocked with fish which provide an extra food supply as well as fertilizing the rice crop. On sloping ground terraces are built into which irrigation water may be led.

Climate has a great influence on the farming system. In most of India there is a distinct rainy season in summer which provides water for the 'kharif' crops (rice, millet and maize) which are sown in July and harvested in October. The 'rabi' crops (wheat, barley and peas) are then sown, and they grow through the warm winter to be harvested in March or April. In some areas it is warm and moist enough in winter for two successive rice crops to be grown in a single year.

Farms are generally small, averaging about one hectare and rarely being larger than four hectares. Also they have often been fragmented by inheritance into a number of separate plots, which causes difficulties in the use of draught animals and ploughs.

In this type of farming much labour is used and very little capital (Fig. 4.1(*c*)). The rice is first sown in nursery beds and later transplanted by hand into the paddy fields. Harvesting is usually done by hand. Productivity per hectare can be very high, so that rural population densities rise as high as 2000 per square kilometre, but productivity per worker is low, so living standards are low.

The Green Revolution has offered the prospect of improved crop yields. In the 1960s the so-called 'miracle rice' was introduced, a variety of rice with an increased yield and a short growing season. Also, new high-yielding varieties of wheat and maize were introduced, but these new varieties need fertilizers and irrigation facilities, so many small farmers have not been able to benefit.

Modern developments in China

Figure 4.12 shows the variations in agriculture in China. Moving from south to north across the country the rainfall total and the length of the frost-free season both decrease. Similarly, the importance of rice decreases from south to north. It is the chief crop south of Shanghai, and in Hainan in the far south three crops per year are obtained. In North China rice is of little importance and wheat, millet and kaoliang take its place.

Before 1949 China's agriculture was typical of the Far Eastern intensive subsistence type. Farms were small and fragmented, and were occupied mostly by tenants who paid high rents (up to 40 % of their annual crop) to absentee landlords who usually lived in the cities. Cultivation was based upon human and animal muscle power, and productivity per worker was very low.

In 1949 the People's Republic of China was established. The absentee landlords were eliminated and their land was distributed to the landless peasants. Since the land was distributed to a large number of peasants, farm holdings became even more fragmented than before. So mutual aid teams were established in which farmers shared their labour, equipment and draught animals. In the 1950s co-operatives were established, consisting of a number of mutual aid teams. So far land remained private property but, with the establishment of a large number of *collective farms*, based upon groups of villages, private ownership of land virtually disappeared. By the end of the 1950s many collective farms had been combined into *people's communes* in which people all worked as wage earners but were permitted to keep small private plots of land.

The above developments took place in the densely populated parts of China, but, in addition, *state farms* were established in relatively remote areas such as Inner Mongolia and Manchuria (Fig. 4.12). These were intended to open up new land to settlement and to develop mechanized methods of farming. The yield per hectare on the state farms has been less than that in the people's communes.

Some progress appears to have been made. Food production has increased but population has also grown. Mechanization is so far little

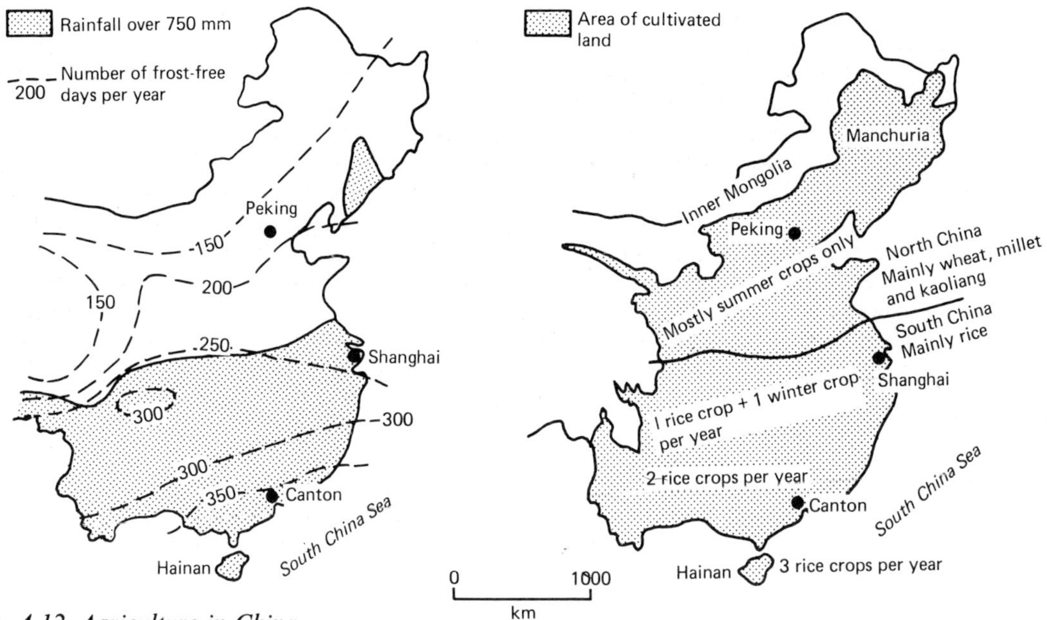

Fig. 4.12 Agriculture in China

developed, but new irrigation works have been created and production of fertilizers has increased. The Chinese have also developed new higher-yielding varieties of rice and wheat.

PLANTATION AGRICULTURE

Plantation agriculture (Fig. 4.3) is one of the chief ways in which the advanced countries of the world have influenced the agriculture of the developing countries. Most plantations are situated in developing countries but they have been established and controlled by people of the advanced countries.

General characteristics of plantation agriculture

A plantation is a large-scale agricultural unit which produces a crop of high value and frequently includes a processing plant for this crop. Commonly the plantation is concerned with the monoculture of a single crop, such as sugar cane, coffee, rubber, tea or bananas, which it exports to world markets. To facilitate this export trade plantations are usually located on or near a coastline.

Plantations represent a highly intensive form of agriculture in the use of both capital and labour. Capital is needed for the initial planting of the crop and the establishment of the processing plant; and labour is needed for the harvesting of the crop. Plantations are usually owned and managed by Europeans, but the labour supply is almost always non-European. They are sometimes regarded in developing countries as undesirable symbols of colonialism.

The evolution of plantation agriculture

In the seventeenth and eighteenth centuries most plantations were located in the American tropics (central America and the northern part of South America). A labour supply could be obtained in the form of slaves from West Africa, and the chief market in Europe was not too far away. A flourishing 'triangular trade' existed between Europe, West Africa and central America.

In the nineteenth century, rapid industrialization in Europe and North America created a demand for food and raw materials. At the same time the abolition of slavery created labour problems. Also the countries of central and South America were gaining political independence with its accompanying disturbances. Hence plantation agriculture began to develop in south-east Asia where a large labour supply already existed. Central Africa was not very attractive because it had not been opened up. However, south-east Asia was much more isolated from European markets until the Suez Canal was opened in 1869.

Many plantations were established in south-east Asia in comparatively sparsely populated areas, so considerable population migrations occurred. These mostly involved the movement of Indians to the rubber plantations of Malaya (Malaysia) and the tea plantations of Ceylon (Sri Lanka). Indians also went to work in plantations in Guyana in South America and in Natal in South Africa.

In modern times subsistence cultivators have frequently taken up the production of plantation crops in south-east Asia. Much rubber is produced in this way in Malaysia.

Examples of plantation crop production

The leading rubber producers are Malaysia and Indonesia, both in south-east Asia, and formerly British and Dutch possessions respectively. The hot, wet equatorial climate is similar to that of Amazonia where the rubber tree originated. Malaysia's labour supply came from India, and Sumatra's (Indonesia) from the nearby densely populated island of Java.

The leading tea producers are India and Sri Lanka. Sri Lanka used to produce coffee in its central highland area but a disease attacked the plantations in the nineteenth century. Tea plantations were developed in the late nineteenth century, using labour brought from India.

Central America (Costa Rica, Honduras, Mexico and Panama) is particularly important for the production of bananas. Special ships, in which the temperature of the fruit is controlled, are used to transport the crop to Europe. Many smallholders grow bananas to sell to the large commercial companies. Bananas are usually produced in hot, wet climates, but they are also grown by irrigation in an arid climate in the Canary Islands.

In Brazil, sugar cane plantations were developed first on the north-east coast, and slaves were imported from Africa. In the nineteenth century coffee was introduced in the south-east of the country and its cultivation expanded inland from Rio de Janeiro on the rich terra rossa soils, using European immigrants as labourers or sharecroppers.

IRRIGATED AGRICULTURE

Irrigation is used for the production of paddy rice even in places where rainfall totals are high, because of the special requirements of the rice plant. In other parts of the Third World irrigation has been developed because of a shortage of rainfall combined with high temperatures and high evaporation rates. Major irrigation schemes such as those in Egypt using the Nile and those in Pakistan using the Indus and its tributaries, are a

An irrigated banana plantation in Tenerife

further example of the way in which the advanced countries of the world have influenced farming in the developing countries.

Problems of irrigation in developing countries

In the Old World arid belt (Fig. 1.2) the basic problem is a shortage of rainfall for crop production. However, rainfall is usually measured as an annual average, and it may vary considerably from this average. Thus, in any particular year, there may be a 50% chance of obtaining a successful crop. This of course allows the possibility of very frequent crop failure.

Even if there is apparently an adequate amount of annual rainfall, it may fall at the wrong time of the year from the point of view of the growing of crops. The rain may come for example at a time when it is too cool to grow crops. If it comes in the hot season, much may be lost through evaporation. Also, much depends upon the type of soil. If the soil is highly permeable, much rain water can be lost through percolation.

Often the most successful irrigation schemes have been based upon a supply of water from streams which rise in well-watered upland areas and then flow across the arid area where irrigation is needed. An outstanding example of this kind of river is the Nile.

In all modern irrigation systems a great amount of capital investment is necessary. In developing countries this capital has usually been provided by advanced countries though, in some cases, notably Iraq, oil revenues have been used to finance the construction of irrigation systems.

Irrigation in the Middle East

Large areas of the Middle East have annual rainfall totals of less than 250 millimetres. In these areas irrigation is essential for successful agriculture (Fig. 4.13).

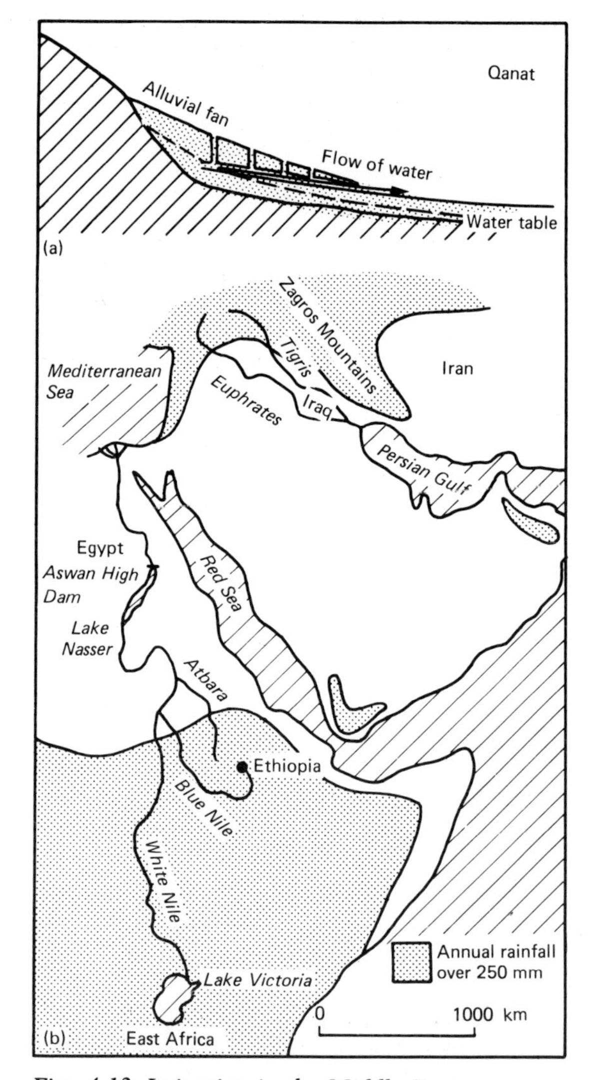

Fig. 4.13 Irrigation in the Middle East

An ancient technique, invented well over 2000 years ago, is the qanat. Qanats can be developed in areas where mountains and plateaux are fringed by alluvial fans. A line of vertical shafts is dug towards the apex of an alluvial fan and a gently sloping tunnel is excavated. When this tunnel reaches the water table in the alluvial fan, water flows along it and can be used to irrigate crops near the outlet of the tunnel (Fig. 4.13(*a*)). The qanat has the advantage of reducing evaporation compared with a surface stream.

In other areas, water may be obtained from wells, either by lowering buckets or, in modern times, by using diesel pumps.

The Nile valley irrigation scheme is worth looking at in detail. In Egypt the Nile flows across a narrow flood plain to the Mediterranean Sea. It rises in the East African highlands where water is plentiful throughout the year, and it collects major tributaries from Ethiopia, where there is heavy rain in summer (Fig. 4.13(*b*)).

Formerly the Nile used to overflow its banks in Egypt in the autumn. The Egyptians built bunds in the flood plain to trap this flood water and allow it to soak into the soil. They were then able to sow wheat, barley and other cool season crops, which could grow through the winter and be harvested in April. Most of the land would then remain fallow until the next autumn floods. This system is known as 'basin irrigation'.

During the nineteenth century the building of dams began, with British assistance. The purpose of these was to store water from the autumn flood and make it available for growing crops during the following summer, thus allowing crops to be grown in summer as well as winter. This is known as 'perennial irrigation'. The most famous dam was the Aswan Dam, first built in 1902.

Since then, with Russian aid, the Aswan High Dam has been built. This allows much more water to be stored in the enormous Lake Nasser. It also protects lower Egypt from flooding and provides sufficient water to extend Egypt's irrigated area considerably.

The Tigris and the Euphrates (Fig. 4.13(*b*)) rise in the wetter areas to the north of Iraq. Their flood season occurs in April and May, just before the very hot, dry summer. Hence they are not so useful as the Nile, but Iraq has used oil revenues to develop irrigation projects.

Much of the cultivated land in Iran is irrigated by streams from the Zagros mountains and from wells. Qanats are now less used than formerly.

4.3 Farming in advanced countries

GENERAL CHARACTERISTICS

In advanced countries farming is not so much a way of life but has become a highly specialized economic activity closely linked with trade and industry. Farming regions tend to specialize in the production of particular combinations of products, mostly for sale to nearby or distant urban areas. Improvements in transport and the development of refrigeration have permitted farms to supply urban markets thousands of kilometres away.

Farming employs only a small proportion of the labour force, but the availability of capital results in a very high level of productivity per worker.

Much farm land in advanced countries is privately owned, but population densities in farming areas are tending to decline as farms are combined into larger units which, by increased mechanization, can develop economies of scale. Very large publicly owned state farms and collective farms exist in the USSR.

Governments by the use of subsidies, grants, loans and restrictions on imports have a great influence upon the type of farming which is carried on. An example of this is the Common Agricultural Policy of the European Economic Community.

Advanced types of farming have spread into the temperate lands of the northern and southern hemispheres with the emigration of Europeans, but little advance has been made into the hot, wet tropical areas.

Frequently farming patterns in advanced countries have characteristics similar to those associated with the von Thünen model (pages 64–70), especially in large countries where the relief is fairly uniform. Broadly concentric patterns of farming types have been recognized in the USA, Australia and Uruguay. Nearest to the major urban markets, farming tends to be highly intensive, with an emphasis on market gardening and dairy farming for milk production. Further away, mixed farming, a combination of crop growing and animal rearing, becomes more common. Farming is least intensive in areas remote from urban markets and usually consists of extensive wheat production or cattle or sheep rearing.

LIVESTOCK RANCHING

Distribution

Livestock ranching (sheep and cattle) earns the lowest net profit per hectare of any major type of farming. It therefore generally occupies remote areas or regions of difficulty where there are few alternative types of land use. In the USA ranches often occupy land which is too dry for crop production. In Britain, extensive sheep rearing often occupies areas of high relief and heavy rainfall.

Livestock ranching generally occupies the semi-arid grasslands of temperate latitudes, especially in western USA, and the interiors of South America, southern Africa and Australia (Fig. 4.3). Immigrants to these areas were Europeans and they adopted European breeds of sheep (the

Merino) and cattle (Aberdeen Angus and Hereford) which are well suited to the temperate climate. Little development has taken place in the tropical grasslands, where pastures are poorer and the climate is unsuited to European breeds.

Characteristics
Livestock rearing in the semi-arid grasslands such as those of the USA began in the nineteenth century as an open-range activity. The land was not divided between different owners, so the ownership of livestock was indicated by branding. In these early days mainly hides and wool, which could be transported easily, were produced. Later, the land was fenced and divided into separate ranches. Water supplies were provided, and railways were built so that meat could become the main product.

In general, livestock ranches are larger than any other kind of farming unit. Compared with the amount of land, very little capital or labour is used, but the amount of capital per worker is quite large; hence labour is very productive. Ranches can be several thousand square kilometres in size.

Although few people work permanently on the ranches, a great deal of labour is needed at certain times of the year, for example when the sheep are being sheared. This problem is solved by having teams of travelling sheep shearers who travel from ranch to ranch.

In the USA most of the cattle are European breeds such as Aberdeen Angus, Herefords and Shorthorns, but crosses with Zebu cattle are used in the warmer south. Irrigation schemes have been developed to provide extra fodder in addition to that obtained from the pastures. Cattle are reared for about two years on the ranches and are then sent for fattening to the farms of the Corn Belt to the east (Fig. 4.14).

In the interior of south-east Australia farmers concentrate more on the grazing of sheep, particularly the Merino breed which produces wool of good quality. Sheep feed on the natural vegetation which is usually of poor quality and may support only about one sheep per hectare. Water is obtained from boreholes (sometimes artesian) and the wool is sent to Sydney, Melbourne and Adelaide for export.

In livestock ranching areas, relatively little processing of the product is carried out before exporting it. The chief exports are live cattle and wool.

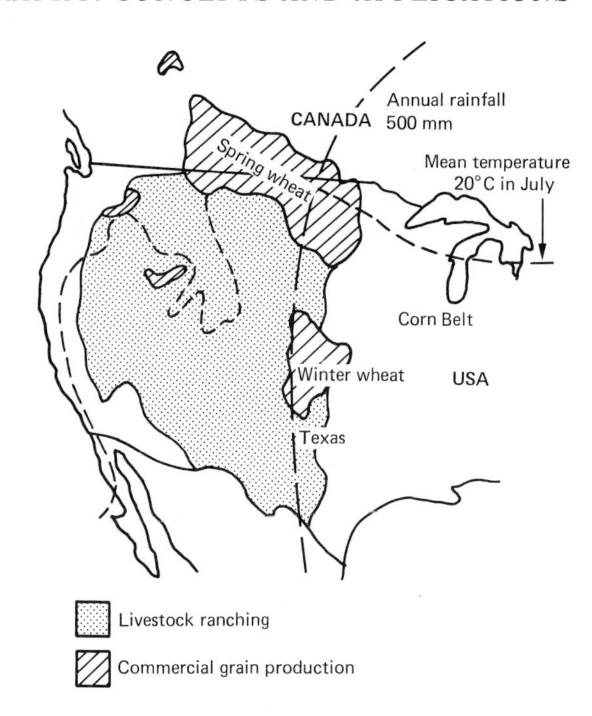

Fig. 4.14 Extensive Farming in North America

COMMERCIAL GRAIN PRODUCTION

Distribution
In the model shown in Figure 4.10 livestock ranching would occupy the outermost ring of all and commercial grain production would occupy the ring immediately inside this and nearer to the urban market. This is because the chief grain crop, wheat, has very low transport costs in relation to its value.

In the mid-nineteenth century, before railways existed, the countries of Europe produced most of their own grain (mainly wheat, barley and oats). In the second half of the nineteenth century railways were built in the USA, Canada, Argentina and south-east Australia, and large-scale production, especially of wheat, began. During the twentieth century, wheat growing has extended into drier and drier areas. In the 1920s it pushed westwards in the USA, and the 'Dust Bowl', which suffered seriously from wind erosion of the soil, was created through the unexpected occurrence of severe droughts. Wheat farmers were forced to retreat from here. In the mid-twentieth century the 'Virgin Lands Programme' in the USSR rep-

Extensive grain production in Canada

resents an extension of grain cultivation into the dry lands of Central Asia (Kazakhstan).

The world's leading producers of the temperate cereals (wheat, barley and oats) are the USSR, the USA, China (pages 73–74) and Canada. Of the countries of western Europe, France is the leading producer.

Characteristics

Farms are usually very large, often over 1000 hectares. Very large state farms have been set up in central Asia. Very little labour is used but each worker is equipped with a large amount of machinery, so his productivity is very high. However, productivity per hectare is low, so land values are relatively low.

Wheat production is strongly influenced by physical factors. Level land is an advantage for the use of large machines for ploughing or harvesting. Approximately 100 frost-free days are needed for the wheat to grow and ripen. Where the winters are relatively mild, winter wheat is sown in the autumn and ripens the following summer. This usually gives a higher yield per hectare than spring wheat which is sown in the spring and grows quickly through a very short growing season. In North America, winter wheat is grown in the south (e.g. Texas) and spring wheat in the north (e.g. the Prairie Provinces of Canada). Hard wheat (for bread) is grown in the drier areas to the west and soft wheat (for biscuits) in the wetter areas to the east.

In North America commercial grain cultivation exists generally on the eastern (wetter) margin of the livestock ranching area (Fig. 4.14). Here the chernozem soils are well suited to wheat growth. In drier areas, wheat is often grown by dry farming methods. These involve growing a crop only every other year so as to store in the soil as much as possible of two successive years' rainfall for the single crop. After the harvest the stubble may be left in the fields so as to trap snow which will provide the soil with extra moisture.

The wheat harvest in North America is often taken by hired bands of farm workers who travel from farm to farm with their machines, beginning with the winter wheat harvest in Texas in May and ending with the spring wheat harvest in Canada in September.

MEDITERRANEAN FARMING

Distribution

A distinctive type of farming has existed for many centuries along the coastlands of the Mediterranean Sea where warm, moist winters alternate with hot, dry summers. Some of the characteristics of this type of farming have been transplanted to central Chile (Fig. 4.3) by emigrant Spaniards, and also to South Africa and parts of Australia. The area in western USA with a Mediterranean climate (central and southern California) has a more commercialized type of farming and is discussed in a later section.

Characteristics

Traditionally farming activities have been closely related to the climate. Cereals, mainly wheat, are grown through the moist winter and are harvested in early summer. Vines and olives are widespread. Their long roots enable them to withstand the summer drought. Flocks of goats and sheep are kept and have traditionally been moved into mountainous areas during the dry summer ('transhumance') but this practice is dying out. In some areas farming is virtually of the subsistence type.

In modern times certain changes have taken place. Citrus fruits have been introduced from eastern Asia, and irrigation schemes have had to be developed to allow them to grow. Spain, Italy and Israel are now leading world producers of oranges and Israel also specializes in grapefruit and Italy in lemons. The coastlands of the Mediterranean Sea are well placed to supply the densely populated areas of north-west Europe with winter vegetables and fruits. Market gardening of this type is less common in the southern hemisphere.

In many areas around the Mediterranean Sea and in Chile farms form part of large estates owned by absentee landlords. This is a serious problem in southern Italy for example.

In general, farming in Mediterranean lands tends to be labour-intensive and living standards are generally low. Little capital is available to improve labour's productivity, except in areas such as Australia where farmers are frequently of north-west European descent.

Detailed land-use patterns sometimes conform to the principles of the von Thünen model which are explained on pages 64–70. Crops that need most attention are grown nearest to the farmhouse or the village so as to reduce the distance farm-workers need to walk to the fields. Hence, in some areas vines may be grown near the village and the more distant fields may be sown with wheat which needs little attention. Also the intensiveness of cultivation tends to decrease with distance from the village. Remote fields may frequently be left fallow.

A traditional farming area in Minorca suffering invasion by urban expansion

MIXED FARMING

Mixed farming areas have a great variety of different farming types. Broadly, mixed farming consists of the production of crops partly for sale and partly for feeding livestock which are then sold (Fig. 4.1(*a*)). But many different kinds of farming exist in mixed farming areas, such as dairy farming (the production of milk and commodities derived from milk), fruit production and market gardening (the production of largely vegetables and flowers). These types of farming are discussed in later sections of this chapter.

Distribution

Figure 4.3 shows that mixed farming has a wide distribution over the world, occurring mainly in areas where European influence has been strong, such as most of Europe, eastern North America and parts of South America, southern Africa, Australia and New Zealand.

Most of the mixed farming areas are favoured with a plentiful amount of rainfall, well distributed through the year.

Many of the world's great urbanized areas are situated within mixed farming areas. These include huge cities such as Buenos Aires, Sydney and Melbourne in the southern hemisphere, and London, Moscow, Paris and the line of great cities extending from Boston to Washington in the USA which has been named 'Megalopolis'. In North America 26 of the 36 cities which had over 1 million inhabitants in 1970 are situated in the mixed farming area of the east (Fig. 4.15) and 20 of these lie in or near the Corn Belt or the Dairy Belt. It appears therefore that mixed farming forms the inner ring of the theoretical land-use pattern of Figure 4.10, nearest to the urban market.

Characteristics

In contrast to livestock ranching and commercial grain production, mixed farming is very diversified. A large number of different crops are grown and these are combined into *crop rotations* in which different crops are grown on a particular piece of land in successive years. This allows soil fertility to be maintained and spreads the farmer's risk over a number of different products.

Large amounts of both capital and labour are applied to the land. This tends to give a high level of productivity both per worker and per hectare. Hence farms are generally smaller than in livestock ranching and extensive grain farming areas.

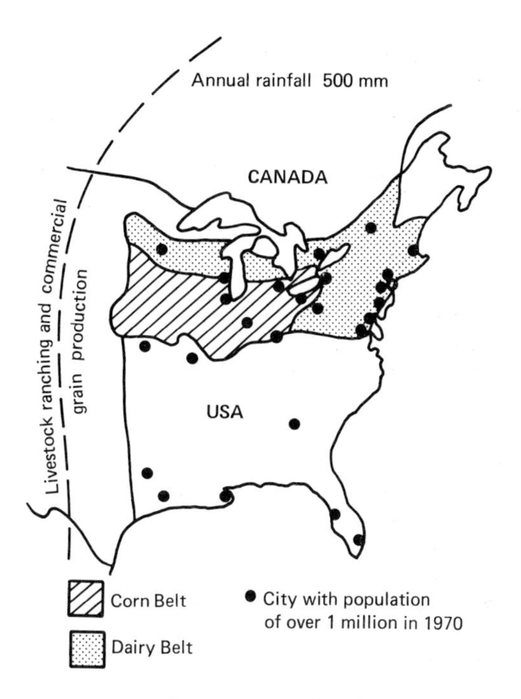

Fig. 4.15 Mixed farming in the USA

Again in contrast to livestock ranching and commercial grain production, very little of the product of mixed farms is exported overseas. Most goes to supply the needs of densely populated parts of the same country. This is particularly true of the USA and Europe, but the more sparsely southern hemisphere countries export rather more.

Farming in the Corn Belt of the USA (Figure 4.15) is based upon the intensive production of maize (corn), but other crops such as winter wheat, oats, soyabeans and alfalfa are included in crop rotations. The area has at least 140 frost-free days and plentiful rain throughout the summer with mean temperatures rising to well over 20°C. The maize crop is largely fed to animals, mostly beef cattle in the west and pigs in the east, nearer to urban markets. Cattle are moved into the Corn Belt for fattening from the ranching areas of the drier parts of western USA. Dairy farming increases in importance eastwards across the Corn Belt, nearer to urban markets. During the twentieth century, as mechanization of farming has increased, farms in the Corn Belt have tended to increase in size, but they are still much smaller than extensive wheat farms or cattle ranches.

DAIRY FARMING

Distribution

Dairy farming has always been very closely associated with towns and cities. Milk is very bulky to transport and soon deteriorates in quality, so it has to be produced fairly near the places where it is to be consumed. Thus, large cities in many different parts of the world and in many different climates usually are surrounded by a ring of dairy farms supplying them with milk. Thus, dairy farming can be regarded as occupying the inner part of the inner ring of land-use surrounding an urban market (Figure 4.10), closer to the city than the type of mixed farming described on page 81 above. Hence, the leading milk-producing countries are the USA, the USSR and highly urbanized European countries such as France, West Germany, the United Kingdom and the Netherlands.

Until the development of refrigerated transport in the 1880s, butter and cheese were also produced mainly near urban markets. With refrigeration it became possible to transport butter and cheese over great distances and the rearing of dairy cattle spread into countries such as New Zealand, remote from urban markets but possessing special climatic advantages for the rearing of dairy cattle. The leading butter and cheese producers are still mainly European countries and the USA and the USSR, but they also include New Zealand and Argentina.

Characteristics

Dairy farming, unlike most other types of farming needs almost continuous attention by farm workers. Twice each day the cattle have to be brought to the farm buildings from their pastures in order to be milked, a process that would be difficult to mechanize. Hence, dairy farms are usually very small in order to minimize the distance the cattle have to be moved, and they are usually owner-operated. The scale of operations could be increased by feeding the cattle in stalls and using a large-scale milking system, but this method is not in common use.

In the Dairy Belt of North America the cool, wet summers encourage the growth of grass, and the hilly relief and stony, glacial soils discourage mechanized crop production. Its eastern part, near the main concentration of cities, specializes in milk production. Further west milk is changed into butter or cheese which have a higher value in relation to their weight and bulk and are therefore cheaper to transport. Canada exports cheese and butter to Britain. The main problem is the long, cold winter, during which the dairy cattle have to be housed and fed indoors. Root crops and maize (for silage) are grown for winter feed. Since continuous milk production necessitates the regular production of calves, these calves are sold for the production of veal.

In the early nineteenth century Denmark was an exporter of grain to other parts of Europe. In the late nineteenth century cheap grain imports began to arrive in Europe as the Canadian Prairies and similar areas were opened up. Denmark therefore began to develop dairy farming. Farms are very small, usually not more than 20 hectares. From the beginning farmers' *co-operatives* were established. The co-operative creamery collects milk from the farms and produces and markets butter and cheese. The skimmed milk is then returned to the farms where it is used to feed pigs, which in turn go to the co-operative bacon factory. The cattle are fed on cultivated crops, especially barley. Though the farms are very small, the co-operatives allow them to enjoy economies of scale, such as the bulk-buying of imported feed or the provision of expensive equipment, and they also control the quality of the product and organize effective advertizing campaigns and marketing techniques.

Most of New Zealand's dairy farms are in North Island where there are rich volcanic soils, a moderate rainfall well distributed through the year, and frost-free winters. European grasses have been introduced to create nutritious pastures. Butter exports really began with the development of refrigerated transport in the 1880s. New Zealand butter has to cross the equator in order to reach Europe. As in Denmark co-operatives have been established, and the skimmed milk is used to feed pigs. The average farm size is about 100 hectares.

FRUIT GROWING AND MARKET GARDENING

Distribution

A great quantity of fruit and vegetables is produced close to urban markets, largely irrespective of climate. Fruit growing and market gardening usually share the inner ring of land use surrounding a city (Fig. 4.10) with dairy farming, but a great quantity of fruit and vegetables is produced in areas fairly remote from cities, which have special advantages of climate.

Different types of fruit tend to be produced in approximately latitudinal belts mostly in the northern hemisphere. In cool temperate climates such as that of Britain, the main types of fruit grown are apples, pears, cherries and plums, together with bush fruits such as raspberries. These can succeed with relatively short, cool summers, and they are dormant through the winter. France, West Germany and Poland, for example, are major producers of apples and pears. Nearer the equator summer temperatures rise and the growing season lengthens. Peaches, apricots and grapes become important and many of the leading producers are situated round the Mediterranean Sea. Citrus fruits need even higher temperatures still and lemons and limes in particular cannot tolerate frost at any time of the year. These are usually produced in the warmest parts of Mediterranean lands. Israel for example is noted for the production of oranges and grapefruit.

North America has sufficient climatic contrasts to produce large quantities of all the types of fruit listed above. Cool temperate fruits are grown in western Canada, in the Great Lakes area and in the northern part of the east coast (Fig. 4.16). Peaches apricots and grapes are grown mainly immediately to the south. The Central Valley of California is well known for its grapes, and Georgia has specialized in peaches. Citrus fruits (oranges, grapefruit, lemons and limes) are largely confined to southern California, Arizona, the Rio Grande delta and Florida, where winter frosts are rare.

Vegetables have a much shorter growing season than most types of fruit. A wide variety of vegetables can succeed in summer in most parts of Europe and the USA and southern Canada, but these vegetables can only be produced in winter if an artificial climate is produced in greenhouses, which adds greatly to production costs. Further south, where the climate is warmer, several successive vegetable crops can be obtained in a single year. Crops such as lettuces, peas, beans and tomatoes, which can only be grown in summer in northern Europe and northern USA, can be produced out of doors in the depth of winter. Tomatoes, for example, are harvested from fields in January in Florida and the Canary Islands.

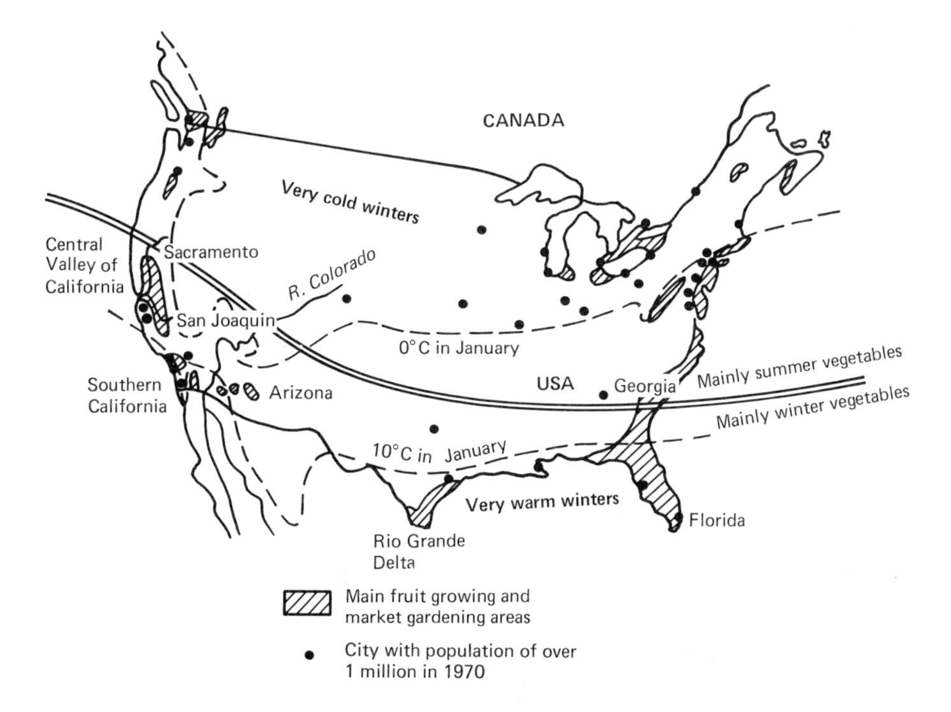

Fig. 4.16 Fruit growing and market gardening in North America

These sell at a price that is high enough to compensate for their transport costs to the cities of higher latitudes. Thus, as shown in Figure 4.16 California, Arizona, the Rio Grande delta and Florida specialize in the winter production of vegetables.

Characteristics

Fruit growing and market gardening use a great deal of labour and capital, so productivity per hectare is high and farms are sometimes quite small. Capital includes greenhouses and irrigation systems where necessary. As in dairy farming, constant attention by the labour force is necessary. Much of the labour used in California consists of immigrant Mexicans. Much fertilizer is applied to the land to ensure high yields of good quality produce.

Fruits and vegetables are frequently very perishable and they need to be delivered quickly to the markets. In modern times, however, the immediate market may be a freezing or a canning factory to which the product needs to be delivered within a few hours of being harvested. Such processing factories allow the market season of summer-producing market gardening areas to be extended throughout the year.

Fruit and vegetable farms are frequently irrigated. Even in humid areas, yields can be improved substantially by the use of sprinkler irrigation in dry spells during the growing season. The irrigated areas in North America shown in Figure 4.3 are largely fruit and vegetable producing areas.

A great irrigation scheme has been developed in the Central Valley of California, known as the *Central Valley Project* (Fig. 4.16). This valley has two major rivers, the Sacramento flowing from the north, and the San Joaquin from the south. In the valley rainfall totals increase steadily from south to north. Hence the northern part of the valley (the Sacramento Basin) has a surplus of water, but the southern part of the valley has a serious water shortage. A canal has been built to transfer water from the Sacramento river to the southern part of the valley. Water from the San Joaquin river is then used to irrigate the far southern end of the Central Valley where no sizeable rivers exist. The irrigated land is used to produce winter vegetables, grapes and, in the warm south, citrus fruits.

Winter vegetables and a wide variety of citrus and other fruits are also produced by irrigation in southern California and Arizona. Irrigation water is obtained from the Colorado river (Fig. 4.16) and two of its east bank tributaries, the Salt and the Gila, in Arizona. Aqueducts carry water from the Colorado river to Los Angeles on the west coast, a distance of about 500 kilometres, and also to the Imperial and Coachella valleys to the east of Los Angeles. In Arizona, farmers in the Salt and Gila valleys are supplied from dams on these rivers. A very large, growing urban market exists for fruit and vegetables in the Los Angeles area where there are four cities with over a million inhabitants. Winter produce is also delivered to the distant cities of the north-east of the United States.

Exercises

1. Discuss the extent to which the von Thünen model provides a satisfactory basis for understanding the spatial distribution of types of farming
(*a*) at the national scale;
(*b*) at the world scale.

2. Illustrating your answer by reference to actual examples, describe and explain the differences between
(*a*) intensive and extensive farming;
(*b*) commercial and subsistence farming.

3. (*a*) Describe the variations in agricultural productivity
(i) per unit area;
(ii) per worker, in different parts of the world.
(*b*) Explain why these variations exist.

4. Explain the ways in which some areas of the world have been able to develop intensive types of farming as a result of the improvements in transport that have taken place in the last century.

5. Evaluate the influence of physical geographical factors on the characteristics (including the world distribution) of the following types of farming:
(*a*) nomadic herding;
(*b*) shifting cultivation;
(*c*) livestock ranching;
(*d*) market gardening.

6. In this question the assumptions and principles of the von Thünen model apply.

(*a*) The table below gives prices and costs of three farm products in an area in which one central city is the only market for farm produce.

	Milk	Wheat	Wool
Market price (per hectare) at the central city	£80	£60	£35
Production cost (per hectare)	£20	£10	£5
Transport cost (per hectare) per 10 kilometres	£10	£5	£2.50

(i) At what distance from the central city is
1. the extensive margin of wool production?
2. the margin of transference between milk and wheat production?
3. the margin of transference between wheat and wool production?

(ii) By how much would the production costs (per hectare) of wheat have to rise before wheat production ceased completely?

(*b*) In the map (Fig. 4.17) A and B are two cities located on a perfectly uniform (isotropic) plain. In the light of the von Thünen model describe and explain the spatial distribution of the three types of farming and the net profits derived from each.

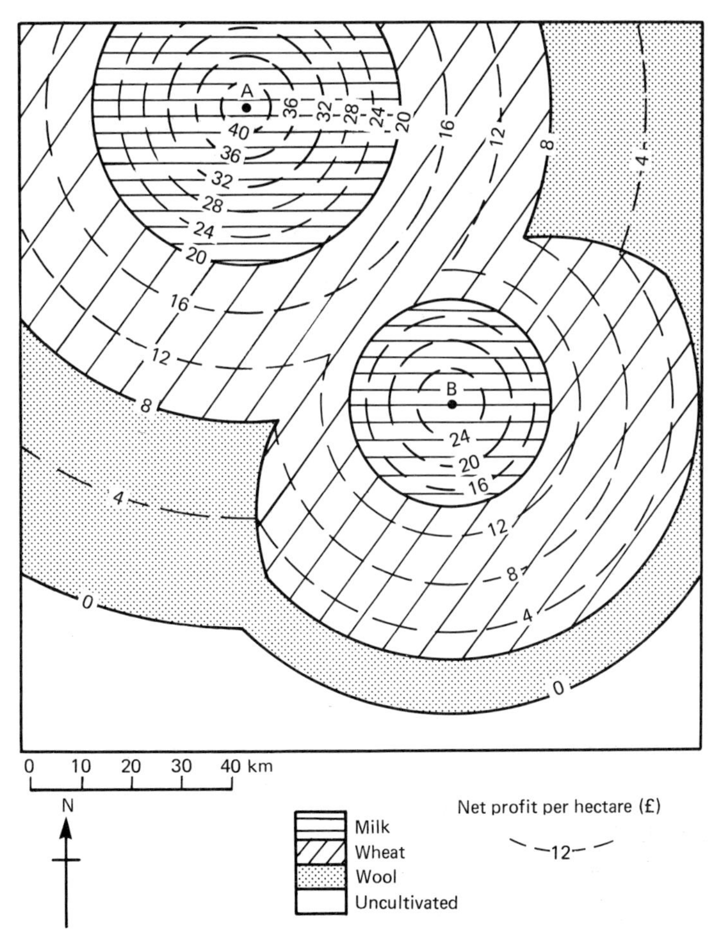

Fig. 4.17

5 Fishing, forestry, power and mineral production

5.1 Fishing

WORLD DISTRIBUTION OF FISHERIES

The leading fishing countries of the world are shown in Figure 5.1. These 16 countries are responsible for catching approximately three-quarters of the world's fish supply.

It is not surprising that very large countries with huge population totals, such as China, the USSR, the USA and India have very large fish catches since they have coastlines of great length and variety. However, many quite small countries, with tiny population totals, have unexpectedly great catches of fish. These smaller countries are grouped in particular parts of the world. Japan, the world's leading fishing country, South Korea, Indonesia, Thailand and the Philippines are grouped, together with China and India, in the Far East. Between them they land at least 40% of world production. The remaining areas are western Europe (chiefly Norway, Denmark and Spain), North America, the west coast of South America (Peru and Chile) and South Africa.

Within the 16 leading producers there is no clear relationship between the size of the fish catch and the total population of the country. Norway (population about four million) for example lands approximately the same amount of fish as the USA with over 200 million people, or even India, with over 600 million. As is shown in Figure 5.1, some relatively small countries have an enormous fish catch in relation to their population. Japan for example produces the equivalent of one tonne of fish for every 10 inhabitants. Norway's ratio is even higher. The USA on the other hand produces only one tonne for every 100 of its inhabitants.

FACTORS INFLUENCING THE DISTRIBUTION OF FISHERIES

Perhaps the most important factor influencing fishery distribution is the distribution of the supply of plankton in the seas of the world. Plankton consists of tiny plants (phytoplankton) and animals (zooplankton) which form the basis of a food chain which supports the fish population. Many kinds of fish (usually small ones) feed on the plankton, then larger fish feed on smaller ones. In some parts of the world's seas plankton is particularly abundant, notably on shallow continental shelves where rivers bring nutrients into the sea, in areas where warm and cold ocean currents meet and mix, and also in coastal areas where there is an upwelling of water from the ocean depths. Plankton is also generally more abundant in high than in low latitudes.

Major fishing areas tend to coincide with these areas of abundant plankton. Thus fishing is concentrated on the continental shelves of north-west Europe, north-east America (near Newfoundland) and eastern Asia. Cool ocean currents and upwelling of water tend to occur on the west coasts of South America (Peru and Chile) and Africa (South Africa). Temperate latitudes provide better

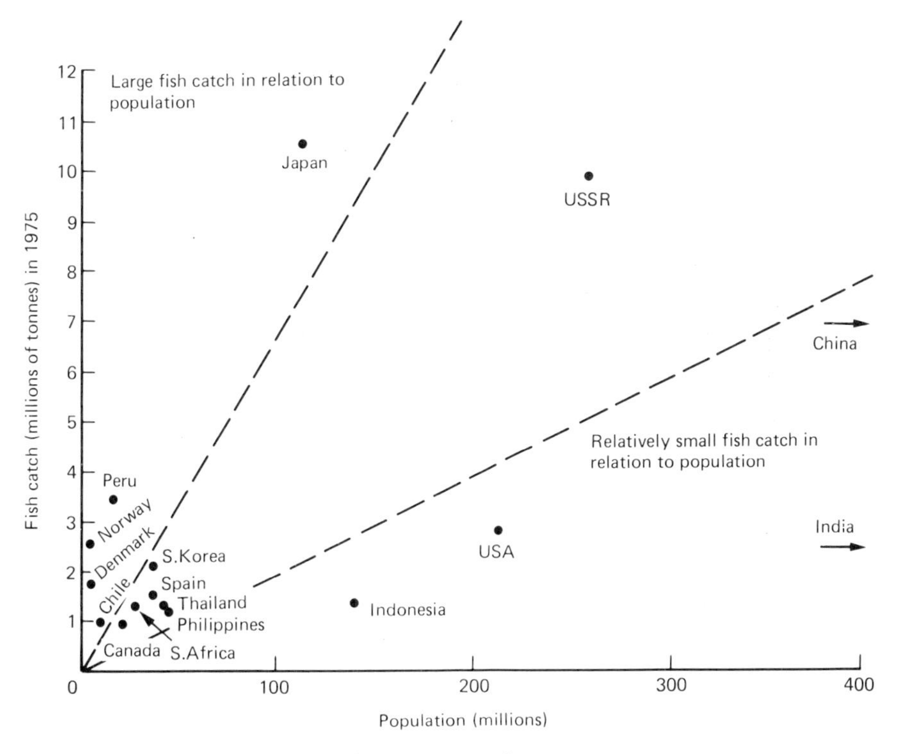

Fig. 5.1 Total fish catch in relation to population

fishing grounds than the tropics partly because they have a greater supply of plankton, but also because there are larger numbers of fewer species of fish in these cooler waters.

If the fish is caught primarily as a fresh food supply, access to well-populated market areas is important. Large populations exist in western Europe, the Far East and north-east North America. However, commercial fishing can take place in quite remote, sparsely populated areas if the fish are canned before being sold (as with salmon in British Columbia) or if they are processed into fish meal, to be fed to animals (as in Peru). The development of factory ships has weakened this link between fishing grounds and their markets. Japan and the USSR have developed fishing fleets which range very widely over the world's oceans.

Fishing near Port Renfrew, British Columbia, Canada

THE FISHING INDUSTRY OF PERU

Before the Second World War Peru was an insignificant producer of fish, but by the 1960s it had become the leading world producer. The cool Peruvian Current and a strong upwelling of cold water from the depths provided ideal conditions for the breeding of plankton upon which fed a huge population of anchovies and other fish.

Commercial fishing concentrated upon the catching of anchovies and processing them to produce fish meal for feeding livestock. This was exported to intensive cattle and poultry rearing areas in many parts of the world. The anchovy population appeared to be able to maintain itself despite intensive fishing. In 1972 production suddenly declined as anchovies became scarcer. Nevertheless Peru remains a leading producer of fish.

POLITICAL ASPECTS OF THE FISHING INDUSTRY

As has been explained on page 34, the fishing industry has been strongly influenced by political factors. During the 1970s, Iceland has gradually expanded its fishing zone from 12 miles (19 km) to 200 miles (320 km) offshore. This has had serious repercussions on the British fishing industry which previously depended greatly upon the use of Icelandic waters by its trawlers. Consequently some British fishing ports, such as Fleetwood, have suffered a serious decline in their trade.

5.2 Forestry

WORLD DISTRIBUTION OF TIMBER PRODUCTION

Practically the whole of the world's supply of timber comes from the three types of forests shown in Figure 5.2: temperate coniferous forests, temperate mixed forests, and tropical broadleaved forests.

Temperate coniferous forests produce softwoods, mainly pine, spruce, fir and larch. They occupy a large area in the north of the northern continents of North America and Eurasia. In the past their southern margin has been pushed back

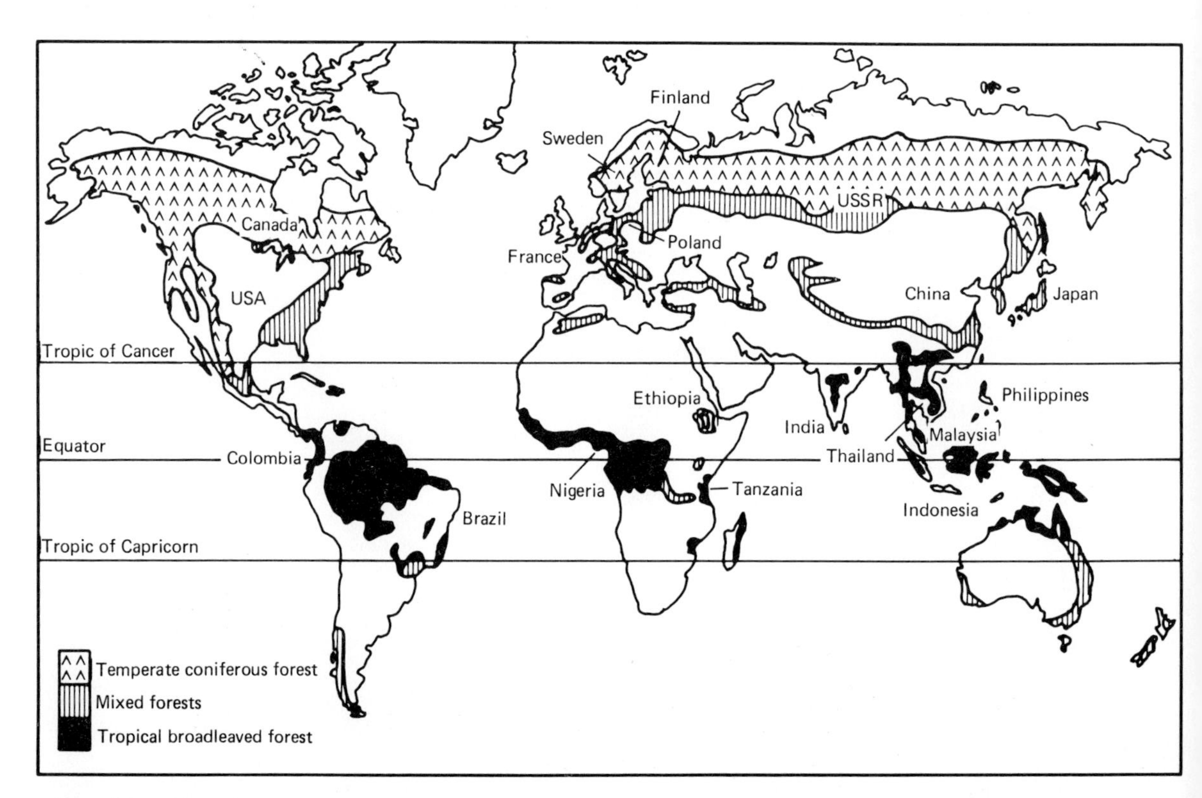

Fig. 5.2 Major forest resources of the world

by the expansion of agriculture, but little further retreat is likely because of the cold climate and the poor, leached soils. Coniferous forests extend southwards along the mountain ranges of western USA and they also exist in Europe in mountain areas such as the Pyrenees and the Alps, and in areas of poor sandy soils such as the Landes of south-western France and the infertile fluvio-glacial deposits of the North European Plain in Germany and Poland.

Temperate mixed forests to the south contain both coniferous softwoods and broadleaved hardwoods such as oak, chestnut, maple and beech. Over many years much of the natural forest here has been destroyed in order to create farmland and here several major concentrations of population have come into being, in eastern North America, Europe and the Far East. A few areas of mixed forests also exist in South America, Australasia and at relatively high altitudes in the Tropics.

Tropical broadleaved forests produce hardwoods such as teak, ebony and mahogany. These forests have not been exploited to the same extent as the forests of temperate latitudes, except in south-east Asia.

The leading countries for the production of coniferous softwoods are the USSR, the USA, Canada and China (Fig. 5.3). They contain enormous areas of forest, but, apart from Canada, their production is not very great in relation to their population totals. Much smaller countries with fewer than 10 million inhabitants, such as Sweden and Finland, are also leading producers. Nevertheless, over 70% of the world's output of coniferous softwoods comes from Canada, the USA, the USSR and China.

The USA, the USSR and China are also major producers of broadleaved hardwoods, these being derived from the large areas of temperate mixed forests which they possess. Canada, on the other hand, lies too far north to contain large areas of broadleaved forest. Other leading producers of broadleaved hardwoods are mostly located in the Tropics (Fig. 5.2). South-east Asia is of particular importance, where Indonesia, India, the Philippines, Malaysia, Thailand and Burma are all leading producers. Elsewhere, Brazil is the leading producer of timber in the Tropics, and, in Africa, Nigeria, Tanzania and Ethiopia are major producers of broadleaved hardwoods.

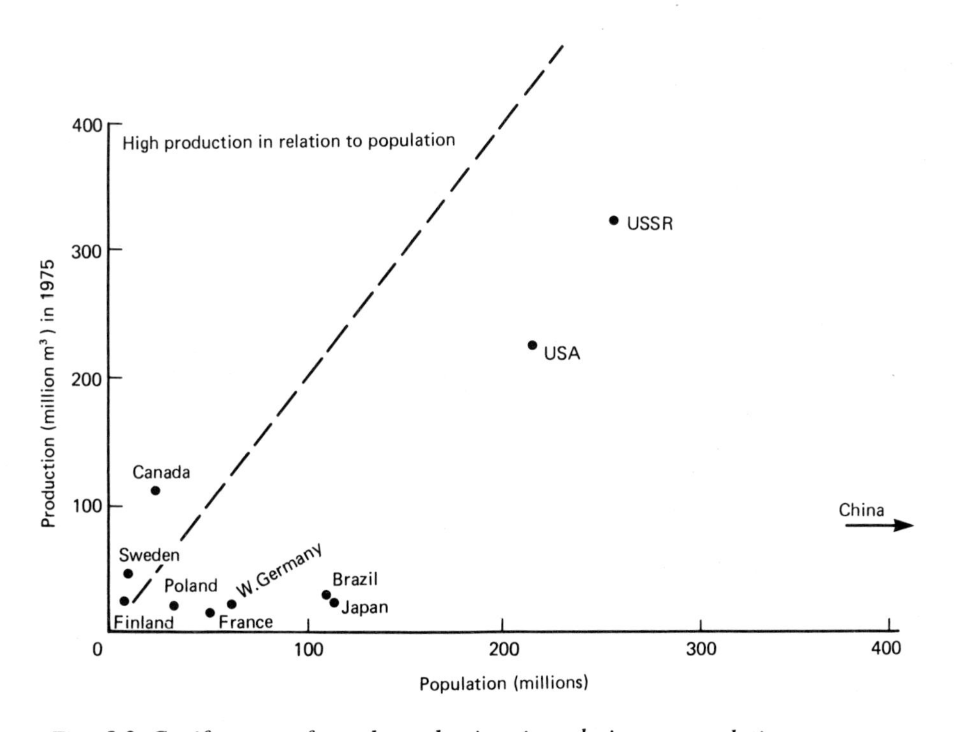

Fig. 5.3 Coniferous softwood production in relation to population

FACTORS INFLUENCING THE DISTRIBUTION OF TIMBER PRODUCTION

Timber is a bulky, relatively low-value commodity. Hence it is an advantage for timber to be produced not too far away from densely populated market areas. Also bulk transport facilities to these market areas are of extreme importance. Sweden, Finland and the Great Lakes area of Canada are relatively near densely populated industrial areas and they have cheap transport routes along rivers and the Great Lakes and the Baltic Sea to these markets. In the USSR most of the production is in the west of the country, where most people live and most industries are located.

The temperate coniferous forests, despite their long, very cold winters when most rivers and lakes freeze, have been easier to exploit than tropical broadleaved forest, because they have fewer species and more uniform stands of single varieties of trees.

Tropical broadleaved forests, on the other hand, have an enormous number of species and it is difficult to concentrate on the production of one type of timber. Also tropical hardwoods tend to be particularly heavy and this has set problems in connection with their transport by river. In addition, there is little demand for hardwood timber in the tropics, so most of the production has to be transported great distances to densely populated, affluent temperate lands. These transport problems are reflected in the location of the main producers of tropical hardwoods. Leading producers in south-east Asia are frequently islands or peninsulas (Indonesia, Malaysia, the Philippines) in which the journey to the coast is comparatively short.

In densely populated areas such as Europe, forestry cannot compete for land with agriculture. The value of its annual output per hectare is much less than that of most types of farming. Nevertheless, areas of low fertility for agriculture may be used for timber production even if they are located near densely populated areas. Much afforestation has taken place in the central mountains of Japan within a few kilometres of major conurbations. Similarly, in Britain, forestry plantations have been established in North Wales, the Pennines and the uplands of Scotland.

FORESTRY IN NORTH AMERICA

Location and types of forest

A broad belt of coniferous forest extends across Canada and the north of the USA from the St. Lawrence valley in the east to British Columbia in the west. Coniferous forests also extend southwards along the mountain ranges that run parallel to the coast of California (Fig. 5.2). Spruce and fir are the chief tree species and particularly large specimens are found along the west coast where the mild climate encourages their growth.

In the eastern USA where the growing season is longer, much of the natural vegetation is broadleaved deciduous forest (Fig. 5.2), but much of this forest has been cleared to create farming land. On the coastal plains bordering the Atlantic Ocean and the Gulf of Mexico there is coniferous forest consisting mainly of varieties of pines. These trees are well suited to the dry, sandy soils and they grow quickly in the warm climate.

Canada has greater forest reserves than the USA but production of timber is greater in the USA. Thus the USA has come to depend more and more upon Canada's forests.

Transport of logs by self-dumping barge near Vancouver Island, British Columbia, Canada

Exploitation of the forests

Exploitation first began in the most accessible places, especially in the eastern USA and in the Great Lakes area. At first most of the timber was used in the sawmilling industry, but later the pulp and paper industry developed and new plantings of trees were made to maintain the supply of pulpwood. The Great Lakes area has benefited greatly from the availability of water transport for timber and the proximity of the large urban centres of north-east USA and their demand for newsprint.

The forest resources of British Columbia and western USA have been exploited more recently and less intensively. Half of Canada's timber reserves are in British Columbia. Sawmills are often sited on coastal inlets such as Puget Sound and the Strait of Georgia. Logs are towed to Vancouver along the sheltered channel between Vancouver Island and the mainland. Sawmilling is so far more important than the pulp and paper industry.

The pulp and paper industry has developed greatly in the southern coniferous forests along the south-east coast of the USA. The supply of pulpwood is quickly renewed through the fast growth of newly planted trees in this warm temperate climate. The large cities of the north-east of the USA provide a market that is not too far away.

5.3 Power supplies

GENERAL CHARACTERISTICS

During the twentieth century there has been an enormous increase in the world's production and consumption of energy. At the beginning of the century coal was the main source of power but, in recent years, there has been a movement towards petroleum and natural gas. In addition, hydroelectricity has increased in importance and nuclear power has been developed.

Logging on Vancouver Island, British Columbia, Canada

The forms of energy used in Britain have changed greatly since 1960. During the 1960s coal made up over 90 % of Britain's energy production, but by the mid-1970s it had fallen to less than 70 % and it is expected to decline even further as its place is taken by oil and natural gas. Nuclear and hydro-electricity have slowly developed but they seem unlikely to make up more than 10 % of Britain's energy production in the foreseeable future.

Most of the world's energy is consumed by a fairly small number of developed countries where living standards are high. Per capita consumption of energy is greatest in North America, the USSR, Australasia and the industrialized countries of northern and central Europe (Fig. 5.4). It is least in the developing countries of central Africa and south-east Asia.

There tends to be a close relationship between countries' living standards and their per capita consumption of energy. In Figure 5.5 the various countries are arranged in almost a straight line across the diagram, showing that energy consumption in general increases as the gross domestic product per capita increases.

COAL

World distribution of coal production

The great majority of the world's coalfields are situated in a belt between latitudes 30° and 60° North stretching across the two great land masses of the northern hemisphere. The USSR and the USA between them produce almost 70 % of the world's coal. Other major producers are China, a number of European countries in which the coalfields are usually located on the northern flanks of the central European uplands, and also Canada, India and Japan. Relatively little coal is mined in the southern hemisphere. Only South Africa and Australia are major producers.

Trends in coal production

Since coal was bulky and difficult to transport, nineteenth century industries tended to cluster on

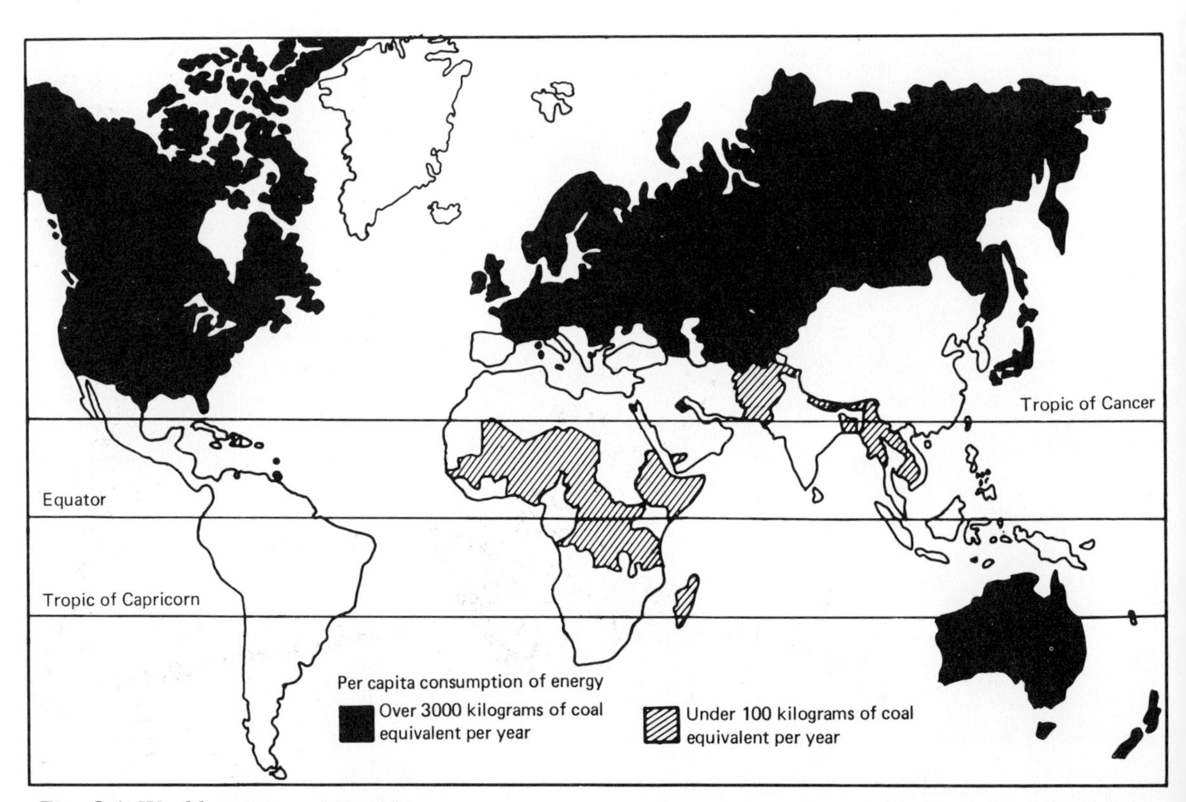

Fig. 5.4 World energy consumption

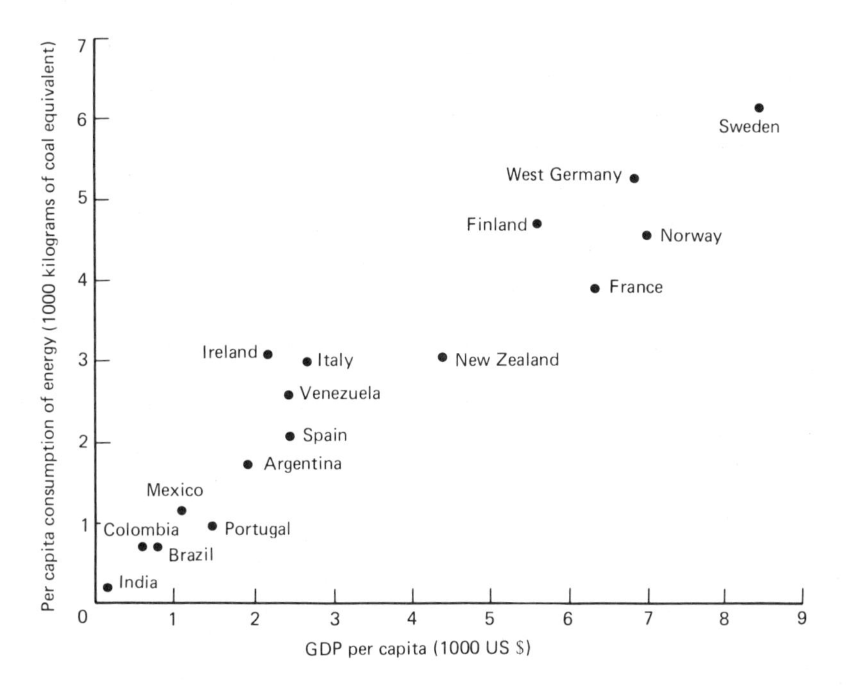

Fig. 5.5 Energy consumption in relation to living standard

the coalfields, particularly in the older industrial countries of Europe. Now, in these countries and in Japan, coal production is declining as competing forms of power have developed. On the other hand, coal production is expanding in the USSR, South Africa and Australia as heavy industries are developed there.

Coal has suffered competition particularly from oil and natural gas in markets such as domestic heating and railway transport. Coal is tending now to be used for the generation of electricity in power stations, the production of coke for metal smelting, and as a raw material in the chemical industry.

Within a single country, coal production has tended to change its location. Usually early coal mines were set up to exploit the most accessible coal seams, usually at a shallow depth on the outcrop sections of coalfields (Fig. 5.6). These seams have frequently become worked out and large, modern collieries have been set up on the concealed coalfields (Fig. 5.6) where the seams are at a greater depth but more plentiful supplies of coal are available. In some countries completely new concealed coalfields have been discovered and exploited in the twentieth century.

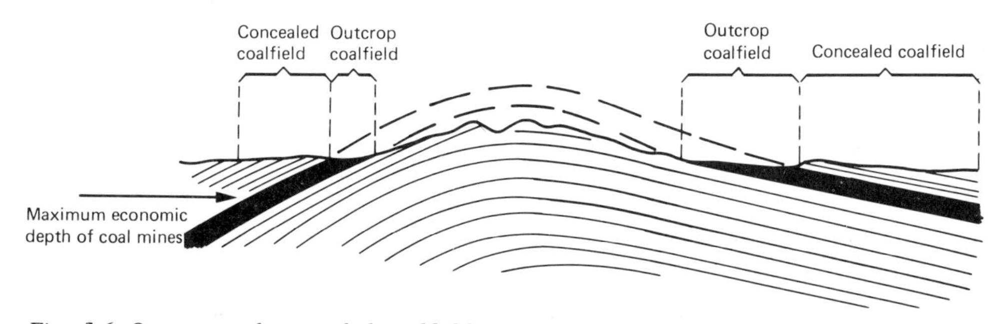

Fig. 5.6 Outcrop and concealed coalfields

Coal production in Great Britain

Most of Great Britain's coalfields are situated on the flanks of upland areas (Fig. 5.7), in the Midland Valley of Scotland, in north and south Wales and to the west and east of the Pennines in Lancashire and Greater Manchester, and in Durham, Yorkshire, Nottinghamshire and Derbyshire. The coalfields flanking the Pennines have large concealed sections where the Coal Measures dip eastwards and westwards away from the Pennine upfold (Fig. 5.6).

Mining is generally most difficult in the coalfields of Scotland and South Wales because the coal seams tend to vary in thickness and they are often faulted and folded. Generally the best mining conditions exist in the Yorkshire–Notts–Derby and the Durham coalfields.

The coal mining industry was nationalized in 1947 and a modernization programme was introduced. During the 1950s coal was easily Britain's chief source of power but imports of petroleum were increasing. In the 1960s natural gas imports developed and the generation of nuclear power increased. By 1970 the coal mining industry supplied less than half of Britain's energy requirements and it had almost been matched by imported petroleum.

The number of active collieries has generally declined, partly because of the decreasing demand for coal and partly because of the reorganization of the industry so as to close down some small uneconomic collieries and to combine others into larger units. In less than 10 years over half of the collieries of Scotland, Wales and north-west England were closed down. The Yorkshire–Notts–Derby coalfield declined least and in 1970 it produced 60% of Britain's coal. New coalfields have been discovered to the east of this coalfield in the Selby area and in the Vale of Belvoir (Fig. 5.7).

The markets for coal have also changed greatly. The manufacture of gas from coal has declined and has been replaced by natural gas from the North Sea fields. Instead of supplying manufacturing industry, domestic users and the railways with coal, the main market has become electricity power stations. A large concentration of these has been developed immediately to the east of the Yorkshire–Notts–Derby coalfield, especially along the river Trent which supplies them with cooling water. Transport of coal to these power stations has been made more efficient by the introduction of 'merry-go-round' coal trains with mechanized loading facilities at the collieries and unloading facilities at the power stations. It is cheaper to produce electricity at very large power stations and transport it to surrounding towns than it is to transport coal to a large number of power stations situated in each town.

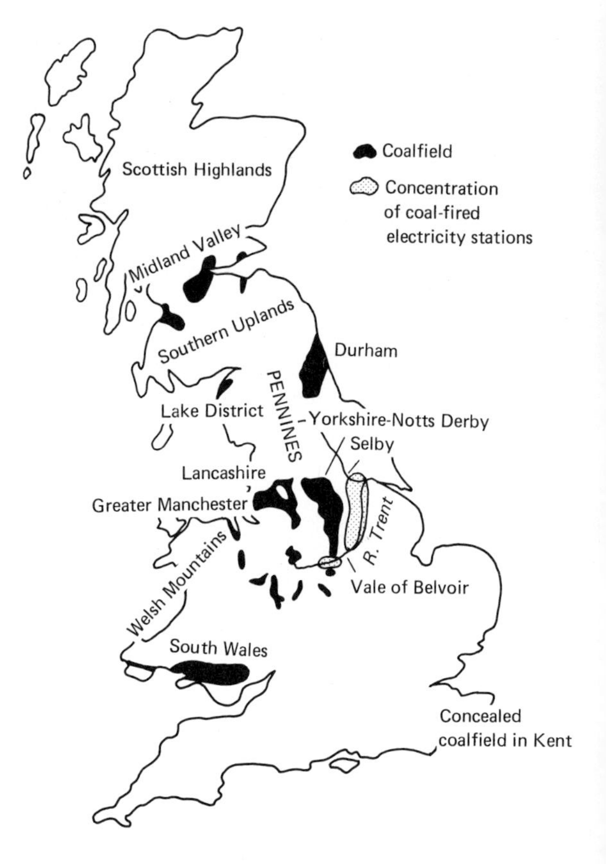

Fig. 5.7 Location of Britain's major coalfields

PETROLEUM AND NATURAL GAS

World distribution of the production of petroleum and natural gas

Petroleum and natural gas fields are commonly found where there are deep accumulations of sedimentary rocks fairly close to fold mountain ranges, so that the sedimentary layers are gently folded so as to provide 'trap' structures. In some

An oil rig in northern Canada

areas, particularly near the Gulf of Mexico, rising salt domes have created 'trap' structures.

In North and South America, oil and gas fields tend to follow the eastern margins of the Rockies and the Andes, both relatively young fold mountains (Fig. 5.8). In the Old World they occur on each side of the Alpine–Himalayan range of fold mountains and also near the Ural Mountains in the USSR. Much of northern Canada, northern Europe, Africa and eastern Siberia are ancient 'shields' with only a shallow cover of sedimentary rocks. Oil and gas fields are rare in these areas.

The world's greatest petroleum producing area lies in and surrounding the Persian Gulf, chiefly in Saudi Arabia, Iran, Iraq, Kuwait and several other small countries. This area also has extremely large reserves. The Middle East normally produces more petroleum than the USA, Canada and the USSR taken together, though the USSR is the leading single producer.

The distribution of the production of natural gas is quite different. Most of the natural gas is produced much closer to the large industrial

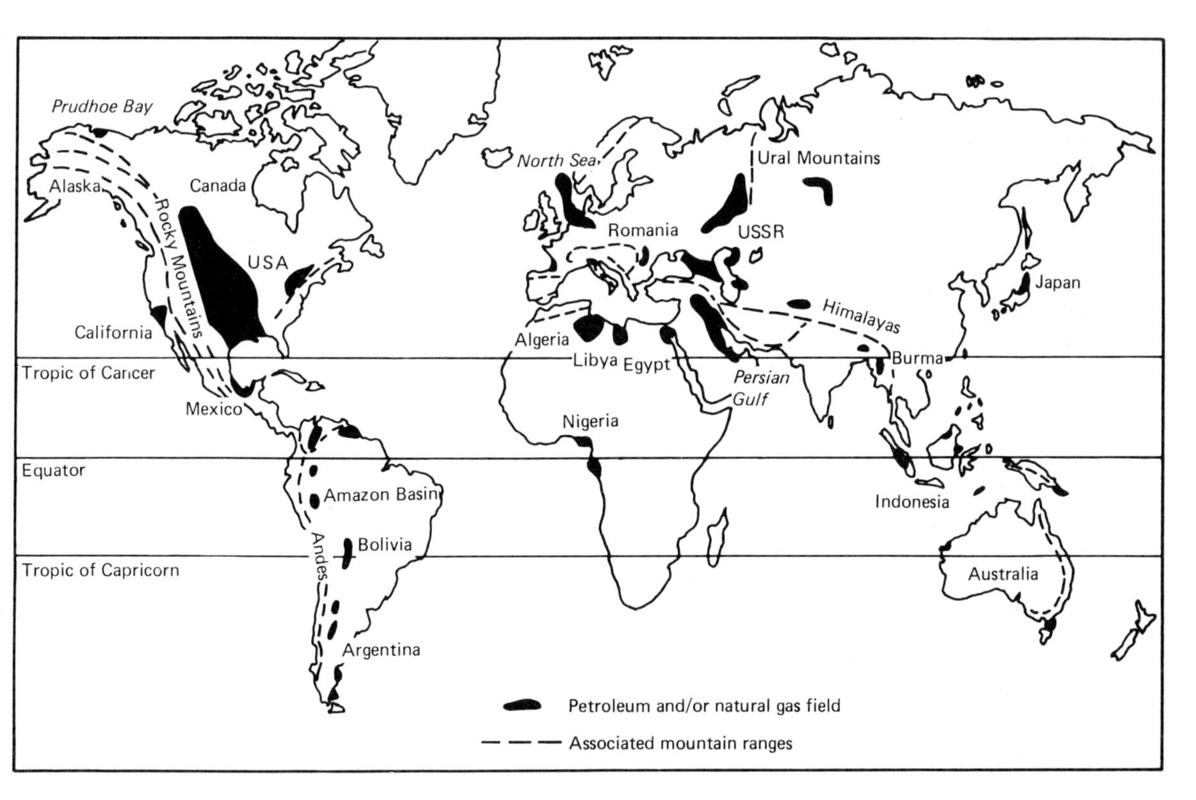

Fig. 5.8 Major petroleum and natural gas fields

centres of the temperate zone of the northern hemisphere. Leading producers include the USA, Canada, the USSR and several European countries such as the Netherlands and the United Kingdom.

Trends in the production of petroleum and natural gas

In recent years the USA has lost its position as the world's leading petroleum producer and the USSR and the Middle East have developed greatly. Europe has become increasingly self-sufficient with the development of the resources underlying the North Sea.

The chief petroleum producing countries of the Middle East together with Venezuela, Ecuador, Nigeria, Gabon and Indonesia formed themselves into the Organization of Petroleum Exporting Countries (OPEC). Because advanced industrial countries had come to depend greatly upon imported oil, the OPEC countries found that they had considerable power in the 1960s and they were able to demand a greater share of the profits of the international oil companies. In 1973 the OPEC countries began to set their own prices for oil and in less than two years they had raised their prices fourfold. The higher prices set by OPEC countries have made it possible for importing countries to develop oilfields in which development costs were previously too great.

Attention has turned especially to exploration of the relatively shallow sea bed of the continental shelves, where drilling costs are several times greater than on land. Submarine oilfields now contribute a large proportion of the world's supply. Also development has taken place in remote areas of the world such as the Amazon Basin and the Arctic coast of Alaska (Prudhoe Bay) which, respectively, are linked by pipeline to the Peruvian coast and the ice-free port of Valdez in southern Alaska. Recently, very large new oilfields have been discovered both ashore and offshore in Mexico.

Production of natural gas has increased very quickly in recent years. Some natural gas is transported by sea in a liquefied form in special tankers but, since most is produced in advanced industrialized countries, much is transported by pipeline direct to consumers. Improved pipeline technology has allowed natural gas to be transported across the bed of the North Sea and the great distance from the Gulf of Mexico to the north-eastern cities of the USA.

Petroleum and natural gas production in the United Kingdom

The large natural gas field at Groningen in the Netherlands was discovered in 1959. Natural gas was known to exist in north Yorkshire. Prospecting therefore took place in the southern North Sea. Successful strikes were made in the mid-1960s especially in the West Sole gas field off the coast of Holderness and in Leman Bank off the Norfolk coast (Fig. 5.9). Most of the gas wells now exist in a broad east-west belt between Lincolnshire and the Netherlands coast. They are often in relatively shallow water. Gas was first piped ashore from the West Sole field to the Easington terminal on the Yorkshire coast in 1967 (Fig. 5.9). Since then the importance of natural gas in Britain has increased rapidly. By the mid-1970s it was contributing nearly 30% of Britain's energy supply.

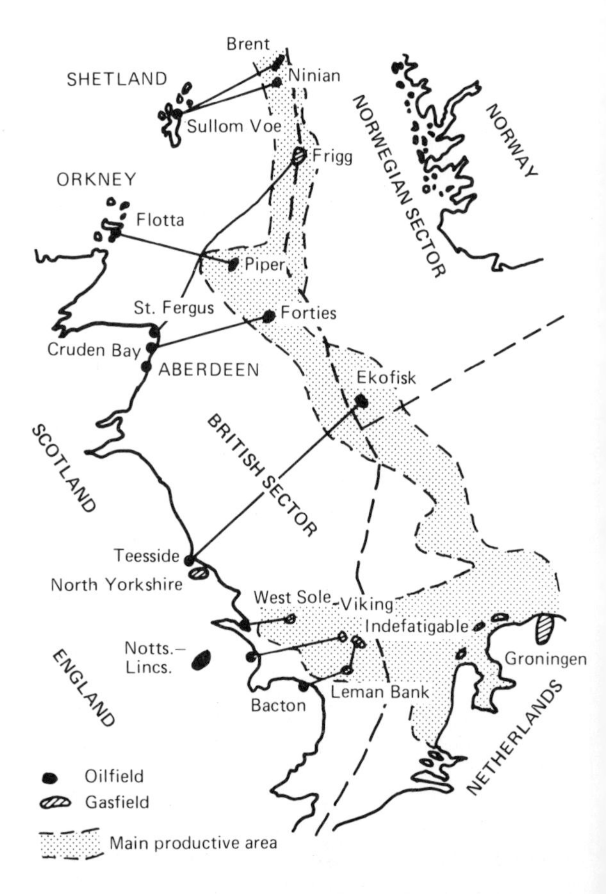

Fig. 5.9 North sea oil and natural gas fields

Before the development of the North Sea oilfields Britain produced very little oil mainly from three small oilfields in England. Most of the North Sea oilfields are situated in the northern part of the North Sea, between Scotland and Norway. Here the sea is much deeper, often over 100 metres deep, and severe storms with waves up to 30 metres high can occur. Nevertheless a large number of wells have been drilled, extending from the Norwegian Ekofisk field opposite central Scotland through the large Forties and Piper fields opposite northern Scotland to the Brent field off the Shetland islands. The Brent field is linked by pipeline to the Sullom Voe terminal in Shetland and the Piper field has a pipeline to Flotta in the Orkneys. A pipeline from the Forties field comes ashore at Cruden Bay, north of Aberdeen and continues to the Grangemouth refinery on the Forth. Oil from the Ekofisk field is piped to Teesside (Fig. 5.9). Natural gas from the Frigg field off the Shetlands is piped ashore at St. Fergus, north of Cruden Bay (Fig. 5.9).

The development of North Sea petroleum has been of great economic benefit to Britain. It has removed the need for Britain to import very large quantities of oil and as a result the balance of payments position has greatly improved. Britain has become a net exporter of oil in the 1980s.

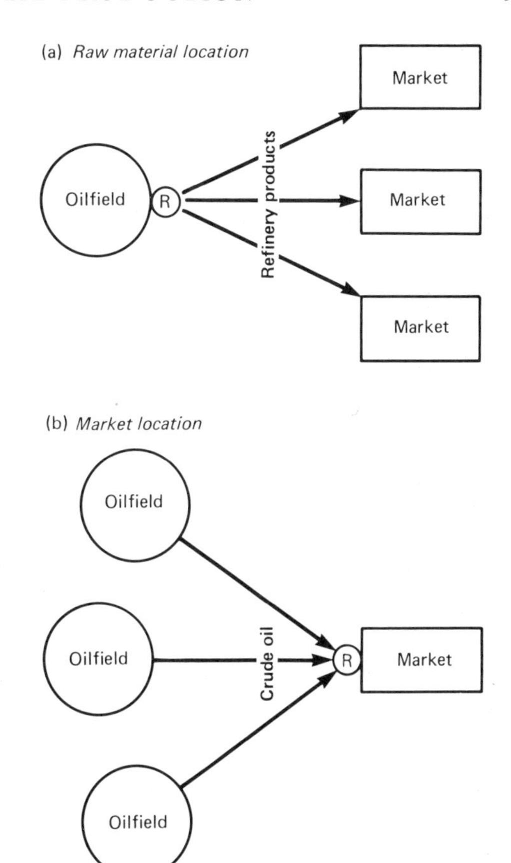

Fig. 5.10 Oil refinery locations

Oil refining

Oil refineries exist to transform crude oil into a variety of so-called refinery products. Chief among these are petrol and various kinds of fuel oil, including fuel for jet aircraft. Oil refineries also supply raw materials for petrochemical industries.

An oil refinery may be located either at the oilfield (the source of raw material) or at the market (where the refinery products are consumed) (Fig. 5.10). If the refinery is located at the oilfield it is necessary to transport the refinery products to the market areas. If the refinery is located at the market it is necessary to transport the crude oil from the oilfield to the market. The location of oil refineries has been strongly influenced by the comparative costs of these two types of transport.

In the 1930s the great majority of oil refineries were located on or near the oilfields, and particularly those of North America and Venezuela, which possessed about three-quarters of the world's refining capacity. Europe possessed very few refineries. Demand for refinery products was still quite small and these could be delivered conveniently in small tankers to consuming areas. At this time, North America consumed the greater part of the refinery products it produced but Europe was supplied mainly from Venezuela and the Middle East.

Since 1945 the oil refining industry has moved strongly towards its markets, particularly towards Europe and Japan and it has declined relatively, though not absolutely, in North America. The USA is the leading oil refining country, but Japan, France, West Germany and Italy are also very important and they import most of their crude oil. The United Kingdom became a leading oil refining country well before the North Sea oilfields were discovered.

Some of the reasons for this change in location are as follows:

(*a*) In the 1930s not all the crude oil was able to be changed into refinery products, and hence there was a loss of weight during the refining process. Thus, it was cheaper to transport the 'concentrated' refinery product than to transport the crude oil which contained a large proportion of 'waste'. The refining process has since been greatly improved, so that there is no loss of weight or bulk at all.

(*b*) Since 1945 it has been discovered that the transport costs of crude oil can be strikingly reduced by the use of very large tankers. A 200 000 tonne tanker can run at less than half the cost (per tonne of oil) of a 25 000 tonne tanker.

(*c*) Since the Second World War there has been much political unrest in oil producing areas, particularly the Middle East, and international oil companies have thought it wise to locate new refineries in advanced countries so as to avoid the risk of nationalization and confiscation of their assets.

(*d*) In advanced countries they could save costs through not having to pay for the provision of an infrastructure of transport facilities, housing and educational institutions.

In the 1970s there are signs that a further change in the location of the oil refining industry is taking place. Some of the underdeveloped OPEC countries, such as Kuwait, Libya and Saudi Arabia, have now become extremely wealthy through the rising price of oil, and it is likely that they will increasingly use their oil reserves as a basis for the development of a variety of industries based upon refinery products. The huge consumption of petroleum products over the world would justify their being transported by very large tankers. The discovery of new oilfields in offshore areas near the British Isles, Japan and other advanced countries means that new raw material sources are being developed near existing refineries. For these reasons the oil refining industry is now tending to become raw material oriented.

THE GENERATION OF ELECTRICITY

General points

There are two main ways of generating electricity. One is to use coal, oil, natural gas, peat or a nuclear reaction to produce steam which powers the electricity generators. Stations that use these methods are known as thermal electricity stations. Heat may also be obtained from the earth's interior. This is known as geothermal power. The other method is to use the power of falling water to drive the generators. This is referred to as hydro-electricity.

Coal-fired electricity stations in Britain are referred to on page 94 and in Figure 5.7.

Thermal electricity stations are usually able to be located in a country relatively near to large populated areas where there is a great demand for electricity. Coal-fired stations are often near densely populated coalfield areas; oil-fired stations near coastal oil terminals; and nuclear electricity stations are frequently located in rural areas on the fringes of large centres of population (Fig. 5.11). This tends to reduce the transmission costs for electricity.

Hydro-electricity stations, on the other hand, must be located at the source of the water power that they use. Hence, the costs of distributing their electricity to markets may be very much greater. Many locations where enormous amounts of water power exist have not been developed because they are too far away from market areas. The operating costs of a hydro-electricity station are very low because there is no need to use any fuel, but their initial capital costs are very great because it is necessary to build dams to create large reservoirs and also to build water channels and sometimes to excavate tunnels so as to collect the water from several drainage basins. The productivity of a hydro-electricity station depends as much on there being a large volume of water as on the water being able to fall from a great height.

An unusual type of hydro-electricity station was opened in the Rance estuary in France in 1966. This uses the energy of the tides.

Thermal electricity stations also have a great need for a supply of water mainly for cooling purposes. Hence they are frequently located on large rivers or on the sea coast. Coal-fired stations use a greater weight of water than of coal, and nuclear power stations use enormous amounts of water.

World distribution of electricity generation

The great majority of the leading countries of the world in electricity generation are the countries of

Europe, together with those parts of the world such as the USA, Canada, South Africa and Australia to which large numbers of Europeans have emigrated. Apart from these, only Japan, India, Brazil, Mexico and Argentina are leading producers and several of these have received many European immigrants. The USA is by far the leading producer, followed by the USSR.

Hydro-electricity is of particular importance in some countries, frequently those with a mountainous relief such as Norway, Sweden, and Austria. Relatively poor countries are usually highly dependent upon hydro-electricity and in some cases, as at the Aswan High Dam in Egypt, the generation of hydro-electricity has been combined with the provision of irrigation water. The two leading producers of hydro-electricity in the world are the USA and Canada, but the relative importance of hydro-electricity to their economies differs greatly. Less than 20 % of the USA's electricity supplies comes from hydro-electricity and most of this is generated in the far west of the country, well away from the major densely populated areas. Hydro-electricity on the other hand contributes about three-quarters of Canada's electricity supplies and most of this is generated in the east, comparatively near to the chief clusters of population.

The USA produces about half of the world's supply of nuclear energy. Other leading producers are the United Kingdom, Japan, West Germany, Canada, Sweden, France, the USSR and Belgium. To a much greater extent than hydro-electricity, nuclear energy is confined to the advanced industrial countries of the world.

Geothermal energy is particularly important in the volcanic areas of the world. Leading producers are the USA, Italy, New Zealand and Japan.

Trends in the generation of electricity

The world production of hydro-electricity has steadily increased over the last 50 years but it still only forms a small proportion of the world's electricity supplies. In the USA, Western Europe and Japan in particular it seems unlikely that sufficient suitable sites exist for a major expansion of hydro-electricity reservoirs, dams and power stations. The development of such projects meets opposition from people and organizations who wish to conserve the environment for recreational purposes. On the other hand, many hydro-electricity projects are being developed in the underdeveloped countries of Africa, South America and south-east Asia.

Between the mid 1960s and the mid 1970s world production of nuclear electricity increased by about 10 times. In the mid 1960s over half of the world's supply was produced by the United Kingdom. Then the other industrialized countries of the world began to build nuclear power stations on a large scale. A great expansion of the production of nuclear electricity took place particularly in the USA, Canada, Japan, and several European countries such as Belgium, West Germany and Sweden. By the mid 1970s the USA was producing about half of the world's supply. Expansion was slower in the United Kingdom, and her share of world production had fallen to less than 10 %. In the late 1970s production extended to less developed countries such as Greece and Iran.

The generation of electricity in the United Kingdom

In the early twentieth century small electricity stations were built in or near towns, almost all of these being coal-fired. During the 1950s oil-fired and nuclear power stations were beginning to be developed and in the 1970s the use of natural gas increased, but coal-fired power stations remain predominant. Great concentrations of these exist along the river Trent and the river Ouse and other

The 2000 MW coal-fired Ferrybridge C power station, West Yorkshire

rivers of Yorkshire (Fig. 5.11), in the London area along the Thames and in central Scotland, south Lancashire and South Wales. They are located near coalfields and supplies of water for cooling, but they are often quite a distance from large towns.

Oil-fired power stations are often located near oil refineries, as at Pembroke (Milford Haven), Fawley (Southampton Water), and the Mersey and Thames estuaries (Fig. 5.11). Fuel can be obtained from the refineries.

Hydro-electricity stations are located mostly in the Scottish Highlands. They have to be located at the source of the water power that they use and they are therefore in remote upland areas, a great distance from markets for the power they generate. Development of hydro-electricity in upland areas near to large conurbations, such as North Wales and the Lake District is restricted because these uplands are National Parks.

Nuclear power stations are distributed widely over Britain (Fig. 5.11). All but one (Trawsfynydd) are situated on the sea coast or a large river estuary because of the need for large quantities of cooling water. Trawsfynydd is situated on the banks of an artificial lake within the Snowdonia National Park. They also need sites where the subsoil is capable of supporting their enormous weight. Most nuclear stations have been built well away from large towns, although as time has gone on this rule has been relaxed and newer stations such as Hartlepool, Heysham and Oldbury are quite near densely populated areas.

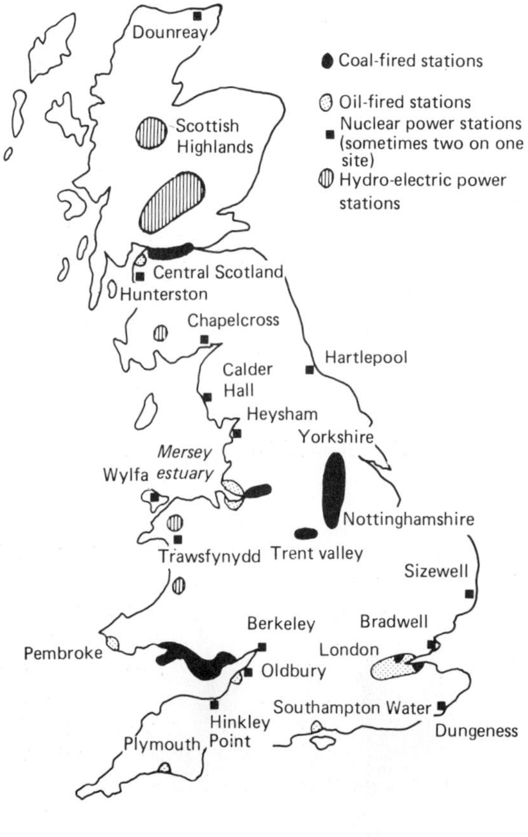

Fig. 5.11 Electricity generation in Britain

5.4 Mineral production

WORLD DISTRIBUTION OF MINERAL PRODUCTION

Many of the world's minerals are mined from the rocks of the ancient shields such as the Laurentian Shield in North America, the Baltic Shield in Europe and the Siberian Shield in Asia. Shields also occur in the southern hemisphere. An ancient continent once existed, which has been named Gondwanaland. Long ago it broke up and the fragments drifted apart to form the Guiana and the Brazilian Highlands in South America, the Western Plateau of Australia and much of Africa. These areas are particularly rich in metallic minerals. Valuable concentrations of minerals also exist in the great ranges of fold mountains that occur on the margins of some of these shields.

Production of some minerals is strongly concentrated in a few places. For example, about three-quarters of the world's supply of gold is mined in South Africa (Fig. 5.12); about two-thirds of the uranium supply comes from the United States and Canada. In a number of other cases, two countries produce over half of the world's supply. These are diamonds (Zaire and the USSR), nickel (Canada and the USSR), manganese (the USSR and South Africa) and

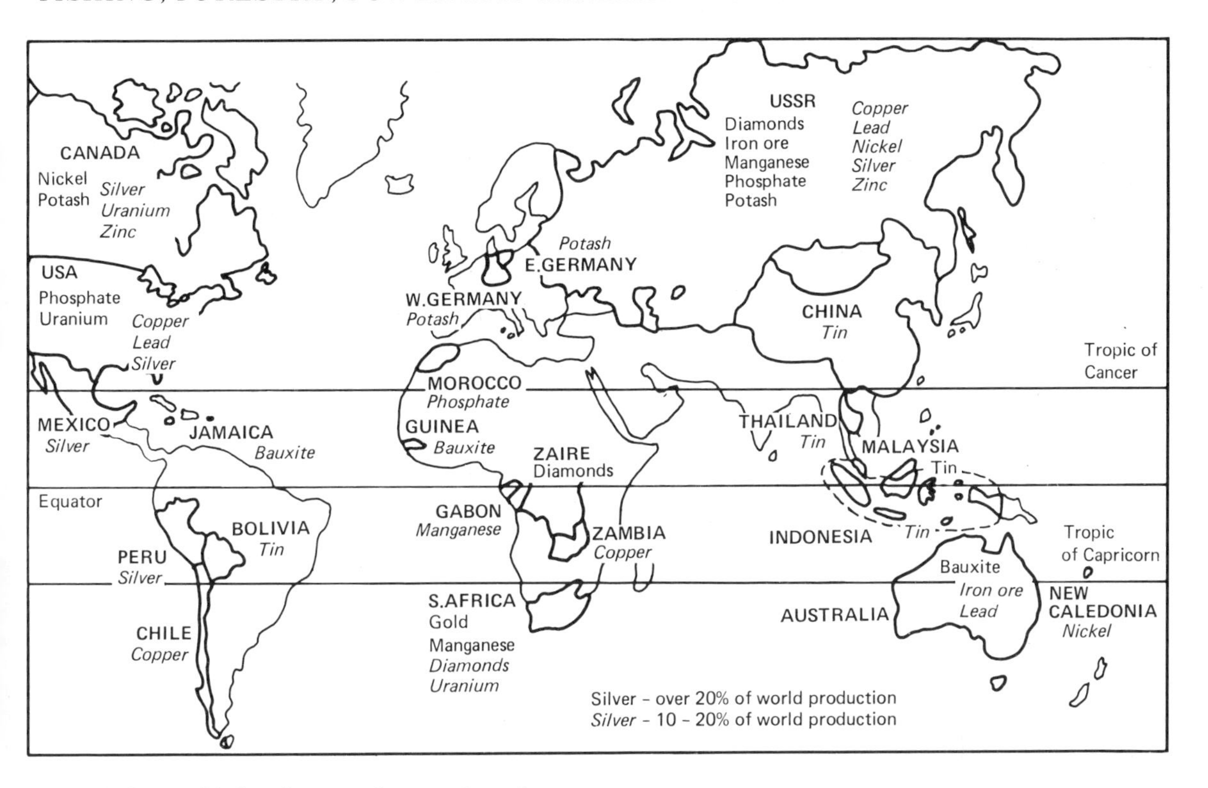

Fig. 5.12 World distribution of mineral production.

phosphate, a mineral used for making fertilizers (the USA and the USSR).

With bauxite, tin and potash (another fertilizer mineral) three countries produce half of the world's supply. Australia produces over a quarter of the world's bauxite, but Guinea (West Africa) and Jamaica are also major producers. Malaysia is the largest producer of tin, but Bolivia and Thailand also produce large amounts. Potash production is dominated by the USSR and Canada, but East Germany is also important. The USSR produces over a quarter of the world's iron ore.

In contrast, other minerals are more evenly distributed and the leading producers do not dominate world production to the same extent. The leading producer of each of the following produces under 20% of world production: copper (USA), lead (USSR), silver (USSR) and zinc (Canada).

Figure 5.12 shows that, in general, the northern hemisphere is more important for mineral production than the southern hemisphere, particularly Canada, the USA and the USSR. This is because the amount of land is smaller in the southern hemisphere and the shield areas are much smaller. Europe is comparatively unimportant as a mineral producer on the world scale.

FACTORS INFLUENCING THE DISTRIBUTION OF MINERAL PRODUCTION

The exploitation of a mineral deposit depends upon three main factors: the mineral content of the rock, the geological conditions in which mining has to take place, and the accessibility of the deposit's location in relation to its markets in advanced industrial countries and sea routes to these countries. In some cases countries develop their own mineral resources so as to avoid being dependent upon foreign supplies.

In extracting an economic mineral it is inevitable that a certain amount of waste rock material is also extracted. If the amount of waste is comparatively great, the mineral deposit is referred to as low grade. Such low grade deposits are frequently worked if they are near to markets. An example is the Lorraine iron field in eastern France, but the development of this field was delayed because its iron ores contained phosphorus and could not be used in the iron and steel industry until 1878. The Kiruna iron ores of northern Sweden have an iron content of 60% and this has encouraged their development despite their isolated location and difficult environment.

Ease of mining is also of great importance. If the mineral is found at a shallow depth, open-cast mining may take place. This is usually cheapter than shaft mining. Malaysia's dominance in the mining of tin is partly due to the fact that its tin deposits exist in alluvial plains. Exploitation is by water pumps and dredges and the heavier tin is then separated from the sediment by gravity methods. Malaysia has been able to compete successfully in British markets with Cornwall because Cornish tin has to be extracted from narrow veins in hard masses of granite.

Transport costs are generally unimportant in the production of high-value minerals such as gold. Their value is so high and their bulk so small that transport costs may almost be ignored. Thus, South Africa is able to produce such a large proportion of the world's gold. Minerals of lower value in relation to their bulk are more strongly influenced by transport costs. Hence production is unlikely to be dominated by a single country. In recent years however bulk carrier ships have been developed that are able to transport low-value minerals great distances at relatively low cost. Particularly iron ore is now transported economically by sea for distances of many thousands of kilometres.

Where large mineral deposits have been discovered in isolated locations special transport facilities have been created to allow them to be developed. The Kiruna iron ore field in Sweden (see above) has been linked by rail to the ice-free Norwegian port of Narvik. A railway over 500 kilometres long links the Schefferville iron field in eastern Canada with the port of Seven Islands near the mouth of the St. Lawrence river. A railway map of either Africa or Australia shows several single routes leading inland from a port. Many of these lines have been built to transport a mineral ore to special loading facilities on the coast.

TRENDS IN MINERAL PRODUCTION

In general there has tended to be a movement of mineral production away from Europe towards the rest of the world. Britain, for example, produced half of the world's copper in the early nineteenth century. Cornwall used to be extremely important for tin mining. As late as 1938 over half of the world's bauxite was produced in Europe.

Development of mineral resources outside Europe has frequently begun with the exploitation of precious metals such as gold and silver and then, with the introduction of cheap bulk transport facilities, lower value minerals have been exploited. A gold rush took place in Australia in the mid-nineteenth century and even as late as 1940 gold made up almost half of the value of Australian mineral exports. Since then Australia has become the world's leading producer of bauxite and is also an important producer of iron ore, lead and zinc. Also nickel and uranium deposits have been developed. Many of these new developments in Australia have taken place in the desert areas of the interior.

In the 1970s there appears to be a distinct tendency for mineral exploitation to extend from the northern temperate zone into the tropics and the southern hemisphere, although the northern temperate zone is still predominant. Discoveries of new mineral deposits are frequently made in developing countries of the tropics such as Nigeria, Zambia, Mexico and Indonesia. Iron ore production has greatly increased in Brazil, where there have also been new discoveries of a variety of minerals including uranium, copper and manganese announced annually.

Exercises

1. Make a reasoned comparison between the world distribution of commercial fisheries and the world distribution of timber production.
2. Quoting regional examples, discuss the importance of an adequate supply of water to the electricity generation industry.

3. To what extent are
(*a*) commercial fishing,
(*b*) timber production
predominantly located in close proximity to the markets for their products?

4. Describe and explain the similarities and contrasts between the world distribution of coal production and petroleum production.

5. Explain the reasons why

(*a*) the oil refining industry has become increasingly market oriented;
(*b*) some types of electricity generating stations are market oriented to a greater extent than others;
(*c*) it is not possible to develop all the world's known power resources.

6. To what extent is world mineral production concentrated in

(*a*) the northern hemisphere?
(*b*) the developing countries?

7. Of the countries shown in Figures 5.13–5.15,

(*a*) which country had the greatest proportional increase
(i) in coal production between 1963 and 1978?
(ii) in iron ore production between 1963 and 1977?
(iii) in crude petroleum production between 1963 and 1977?
(*b*) discuss the extent to which
(i) coal production has tended to shift away from the older industrial countries;
(ii) crude petroleum production has tended to shift from the western hemisphere to the Middle East;
(iii) iron ore production has tended to shift towards the southern hemisphere.

Data adapted from 1966 and 1980 issues of The Geographical Digest, Geo Philip and Son Ltd.

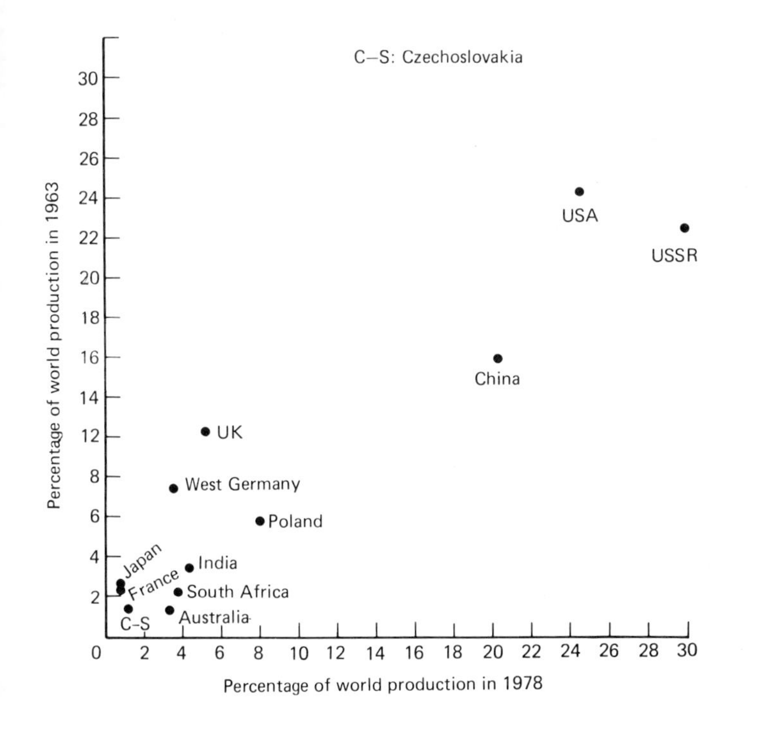

Fig. 5.13 Major producers of coal in 1963 and 1978
Total world production of coal increased by 1.25 times between 1963 and 1978. Only those countries which produced at least 1 % of world production in 1963 are shown on this graph. Canada produced 1.06 % of world production in 1978.

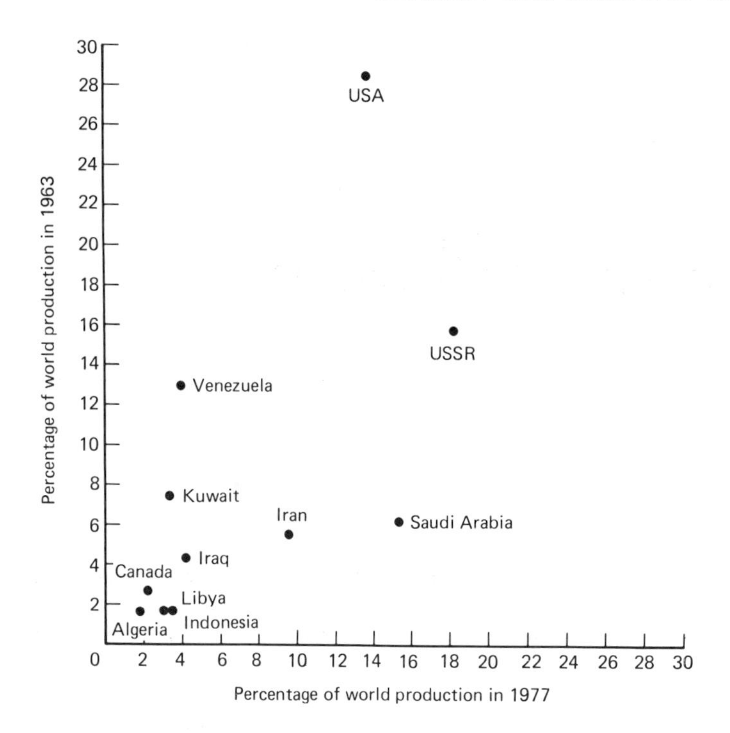

Fig. 5.14 Major producers of crude petroleum in 1963 and 1977
Total world production of crude petroleum increased by 2.3 times between 1963 and 1977. Only those countries which produced at least 1.7% of world production in 1963 are shown on this graph. By 1977, the following had also become major producers: Nigeria (3.4%), China (3.3%) and the United Kingdom (1.3%)

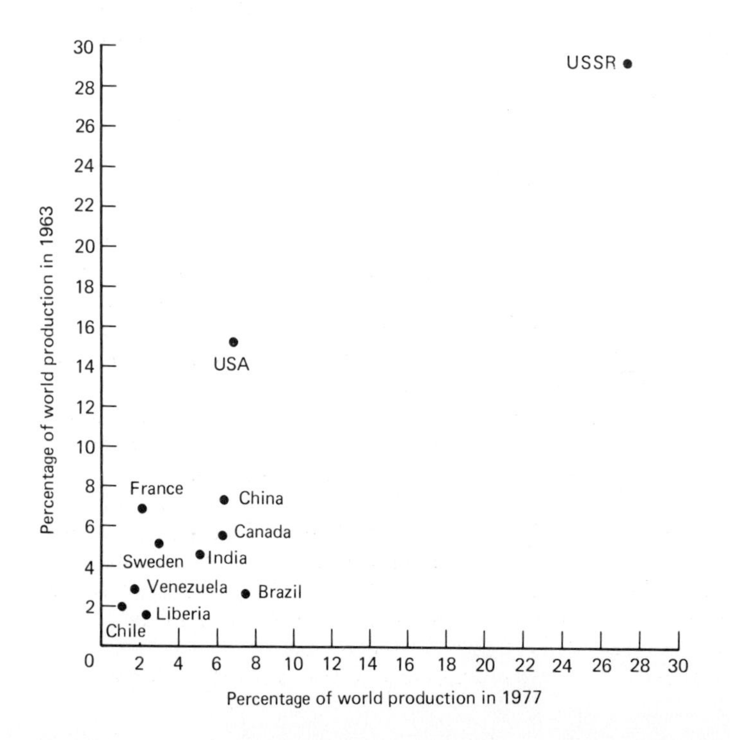

Fig. 5.15 Major producers of iron ore in 1963 and 1977
Total world production of iron ore increased by 1.95 times between 1963 and 1977. Only those countries which produced at least 1.6% of world production in 1963 are shown on this graph. By 1977, the following had also become major producers: Australia (11.6%), South Africa (3.0%) and Mauritania (0.9%)

6 Manufacturing industry

6.1 Factors influencing the location of manufacturing industry

The farmlands, fisheries, forests and mines described in Chapters 4 and 5 produce materials which usually need to be changed in some way before they are suitable to be used or consumed by people. For example, wheat may be changed into bread, fish may be put into cans, timber may be changed into paper, iron ore may be used to produce steel which may eventually may be used to make motor cars. All these are examples of manufacturing industries. In a manufacturing industry raw materials are changed into manufactured products by the use of power supplies, labour and machines (capital) and these products are delivered to markets (often large cities and conurbations) where they are sold. Decisions have to be made about the location of each industry. It may be best for the industry to be located near certain raw materials, or near a particular power supply, or near a supply of labour or capital, or near where the product is sold. In some cases an industry may benefit from being located near to other industries with which it is linked in various ways, thus obtaining benefits from agglomeration. In most countries the government has a strong influence on the location of industry. Figure 6.1 illustrates some of the major factors of industrial location.

It is usually quite difficult to explain why an industry is located in a particular area because it is rare for only one of these factors to be responsible. Usually the location of an industry has been influenced by several of these factors simultaneously. In some cases explanation is even more difficult because an industry may exist in an area as a result of factors that exerted an influence 50 or 100 years ago and now have lost their importance. Such a 'relic' industry is an example of 'geographical inertia'.

THE INFLUENCE OF RAW MATERIALS

Manufacturing industry tends to be attracted towards its raw materials if there is a great loss in weight during manufacture. In the manufacture of butter, for example, the weight of butter produced is only about one-sixth of the weight of the milk used. Similarly, the wood pulp and newsprint manufacturing industries tend to be located near the forests from which their raw material comes. Leading world producers of milk such as the USSR, the USA, France, West Germany and Poland, are also leading producers of butter and cheese. Leading world producers of coniferous timber such as the USSR, the USA, Canada, China and Sweden, are also leading producers of wood pulp and newsprint.

Many ores have a very low metal content. Some

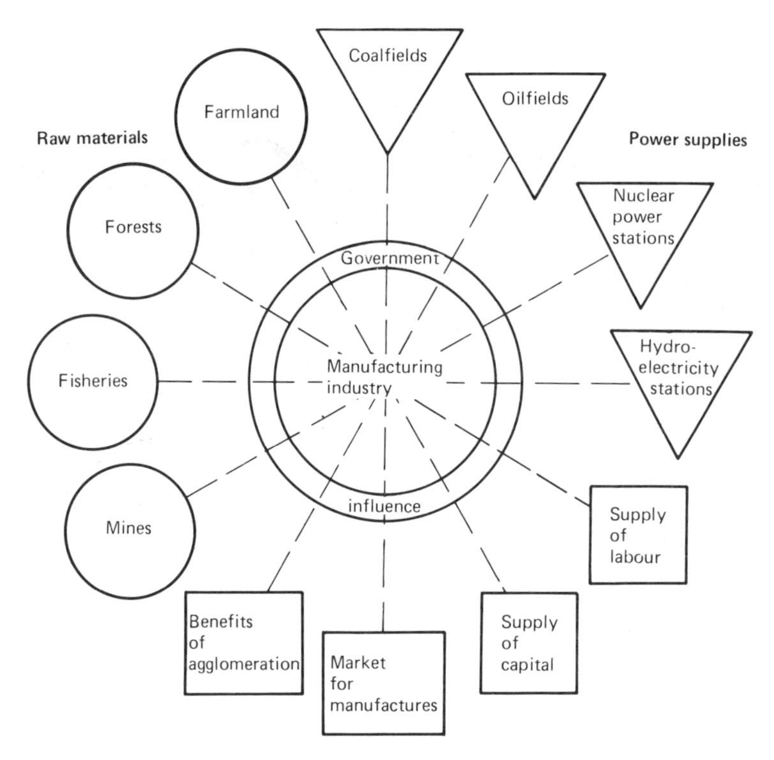

Fig. 6.1 Factors influencing industrial location

copper ores, for example, contain only about 1% of copper. It would be uneconomic to transport up to 99% of waste material, so such ores are processed near the mines in order to remove some of this waste. Some iron ores are of low quality and are beneficiated before being transported away from the mining area. The low quality taconite iron ores in the Lake Superior area of North America have about half of their waste content removed before they are transported.

Food processing industries are commonly located near the supply of raw materials. The manufacture of beet sugar in Europe is generally located in the areas where sugar beet is grown. Raw sugar makes up only about one-eighth of the weight of the raw materials. Sugar cane is also processed into raw sugar in the areas in which it is grown.

Frequently it is necessary to process a raw material in its producing area because it is perishable and would rapidly deteriorate if it were transported. Thus freezing and canning plants for vegetables and fruit must be located in the growing areas. There is also a considerable loss of bulk during processing as for example in the transformation of peas in the pod to packets of frozen peas.

Not all farm products however are processed near their producing areas. Wheat for example is easier to transport in an unmilled state than in the form of flour. Hence flour milling tends to be located nearer to markets and well away from wheat producing areas.

In general as time has passed raw materials have tended to lose their attraction for manufacturing industries. This is because of the introduction of more efficient forms of transport and bulk handling. Also improvements in processing techniques mean that less waste is produced in the manufacturing process.

In many modern industries raw materials have very little influence upon their location. These industries commonly use materials that lose no weight during processing and that are not perishable. The manufacturing process often results in an increase in bulk. Such industries are the assembly of motor cars and electrical goods.

THE INFLUENCE OF POWER SUPPLIES

Over the last two or three centuries supplies of energy have changed greatly. For most of the eighteenth century and in many areas throughout the nineteenth century the main source of industrial power was water power, generated by waterwheels. This was succeeded by coal and the steam engine. In modern times oil and natural gas have become more important, and electricity, generated at coal-fired, gas-fired, oil-fired, nuclear and hydro-electricity stations has become the most important energy supply of all. This sequence of power supplies represents a gradual increase in the mobility of energy so that it has become less necessary for industries to locate at the source of the power supply that they use.

At first, power was quite immobile. A waterwheel could only deliver power by systems of belts and pulleys to machines within a single building. Thus factories tended to be located at suitable sites along the banks of rivers and streams.

In many areas in the eighteenth century coal replaced water power, and in the nineteenth century coal was easily the most important power supply. Coal is bulky and, at first, it was difficult to transport, so many industries were attracted to the coalfields. As time went by, improved transport facilities were created, and coal could be transported along canals, railways and by sea. Thus industries were able to grow near such transport routes.

In the twentieth century oil has increased greatly in importance. It is much easier to transport than coal, by means of pipelines and large tankers. It also contains more heat per unit weight than coal. Hence it has tended not to attract industries to the same extent as coal did. Oil can be delivered comparatively easily to industries located near to their raw materials or on coalfields. Some industries are often located near to supplies of oil, but these usually use oil primarily as a raw material, as in the case of petrochemical industries. The use of oil has given industry a much freer choice of location.

The most mobile and flexible source of industrial power is now electricity, which can now be transmitted over great distances with little loss. In most advanced countries it is available almost everywhere. Thus industries that use electricity have few restrictions on their location from the point of view of power supplies.

An old textile mill complex which formerly used water power and coal, and is now occupied by modern industry

Therefore more mobile types of power supplies such as oil and electricity have challenged the dominance of coal during the twentieth century. It has been possible to deliver these mobile forms of energy to old industrial centres on or near the coalfields. Thus coalfield areas, even in the late twentieth century, remain major centres of industry. This is described as 'geographical inertia'.

In most modern industries the cost of power is only a small proportion of their total production costs, so these industries' locations are not strongly influenced by the availability of power supplies. On the other hand, some industries such as petrochemicals, wood pulp and metal refining, and particularly aluminium refining, need very large amounts of power. Such industries are frequently located where hydro-electricity stations provide abundant cheap energy, even if this involves occupying remote locations such as northern Scotland or Kitimat in northern British Columbia. The world's leading producing countries for electrical energy are also, almost without exception, the leading countries for the refining of aluminium.

The iron and steel industry has changed its location several times in response to changes in its energy supply. Early iron industries used charcoal to produce heat, so they were often attracted by forested areas such as the Weald in south-east

England. In the late eighteenth century the invention of coke smelting resulted in blast furnaces being set up on coalfields where frequently iron ore could be found as well as coking coal. Also the invention of the steam engine further encouraged the movement of the iron industry towards coalfields and away from water power sites. In these early days, coal was used inefficiently: very large amounts of coke were needed to produce relatively small amounts of pig iron. As time went on, improvements in technology decreased the amount of coke which needed to be used, so the attraction of the coalfields became much less. In modern times therefore the iron and steel industry has tended to move away from the coalfields. In Europe particularly, many new iron and steel industries are situated on the coastline where both coal and iron ore can be transported by sea. Older iron and steel industries still exist in coalfield areas as examples of geographical inertia.

THE INFLUENCE OF MARKETS

Manufacturing industry tends to be attracted towards its markets if the product is heavier, bulkier or more fragile than the raw materials. In this case it is cheaper to transport the raw materials a considerable distance to the factory than to transport the product a considerable distance to the market. Many modern industries are of this type, especially the 'assembly' industries in which a variety of different products are brought together at the factory and are made up into a bulkier product, which may also be rather fragile. A good example is the motor car industry in which assembly plants are often located near large markets but components such as body shells, engines and transmission systems may be manufactured elsewhere. This principle also applies to the manufacture of radio and television sets, boxes and other containers and furniture. The leading producers of motor cars and radio and television

The British Leyland car factory at Longbridge, Birmingham – an example of an 'assembly' industry

sets are almost all industrialized countries with high living standards where there is a large demand for these products. These industries are dominated by the USA, Japan, Italy and the countries of north-west Europe.

Some industries such as brewing and soft drink manufacture also produce products that are heavier and bulkier than their raw materials mainly because the manufacturing process involves the addition of large amounts of water to relatively small amounts of other raw materials. The products may then be bottled or canned, thus causing a further increase in bulk. These too are usually located near their markets.

Where perishable raw materials are processed into less perishable products (page 106), a raw material location is often preferred. A market location is usually preferred for industries that transform less perishable raw materials into perishable products. A good example is the baking of bread, a commodity that must be delivered quickly to shops. A rather similar example is the printing of newspapers. The news that these publications convey to their readers is extremely 'perishable' with the passage of time. Most large towns have their own local newspapers which deal largely with local news. The offices that produce them are situated not only near their markets (the readers) but also near their raw materials (local events), but of course the paper they use may have been transported a considerable distance.

Some industries, such as the manufacture of clothing, need to have close contacts with the retail trade and other service industries, which are strongly oriented towards large centres of population. In some cases, such as in clothing manufacture and book publishing, a location in or near a large city may be valued because of the prestige it is thought to give.

Many modern industries have close 'linkages' with other industries, often supplying them with materials or receiving components from them. In large urban areas there is usually a wide variety of industries that are dependent upon one another in these ways. Such industries find it convenient to be near complementary industries and therefore locate in urbanized areas in which they also find their market. The manufacture of textile machinery is frequently concentrated in areas where textiles are manufactured.

THE INFLUENCE OF TRANSPORT COSTS

The location of some industries is strongly influenced by the costs of transporting their raw materials and their products, particularly if both their raw materials and their products are bulky and of relatively low value.

The costs of transporting either raw materials or manufactured products include two elements:

(*a*) The loading and unloading of goods on and off a transport system involves the use of various types of equipment such as docks, warehouses, cranes, grain elevators, etc. A charge is made for the use of these facilities. Such costs are known as *terminal costs*. They do not vary according to the distance the goods are being transported.

(*b*) The costs of actually moving goods from place to place are known as *line-haul costs*. These costs increase as the distance the goods are transported increases. Commonly however they increase at a decreasing rate so that the line-haul cost per kilometre is less for long journeys than for short journeys. Such costs are said to 'taper' with distance.

Figure 6.2(*a*) shows the total costs of transporting a raw material from its source (R) to a market (M) and to any location between the two. The terminal cost (the cost of loading the raw material at its source and of unloading it at its destination) is represented by RA. This is a fixed cost regardless of transport distance. The slope of line AP represents the tapering increase in line-haul costs as transport distance increases. Point X is situated midway between R and M. It will be noted that transport costs increase more between R and X than between X and M. This is because of the tapering of the line-haul costs.

Figure 6.2(*b*) shows the total cost of transporting a manufactured product to the market (M) from R or from any location between the two. The terminal cost (the cost of loading the product at a factory and of unloading it at the market) is represented by MQ. RB is the total cost of transporting the product to the market (M) if the factory were located at the raw material source (R). It will be noted that this total transport cost is very little greater at R than at X because of the tapering of the line-haul costs.

Figure 6.2(*c*) shows the total costs of transporting the raw material (AP) and the manufactured

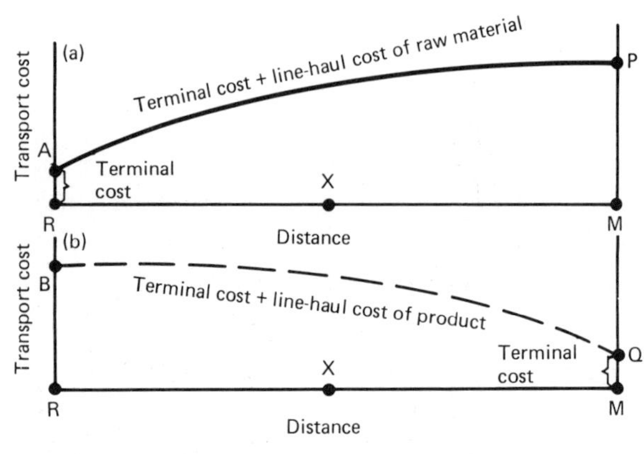

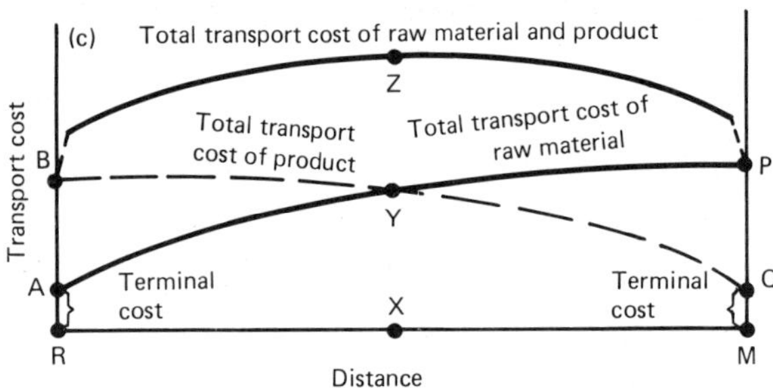

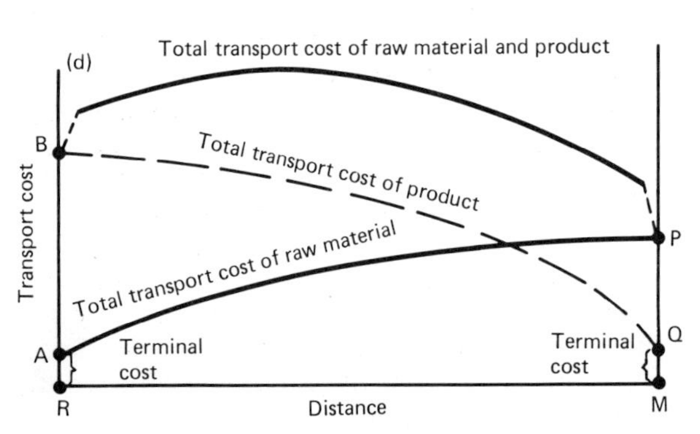

Fig. 6.2 Influence of transport costs upon industrial location

product (BQ). If a factory were to be located at X, its total transport costs would be XY to bring the raw material from R, plus XY to deliver the product to the market (M). Total transport costs at X (XZ) are greater than anywhere else. The total transport costs at R are only RB (i.e. terminal and line-haul costs of transporting the product to the market). The raw material need not be transported at all, so line AP can be ignored. Correspondingly the total transport costs at the market (M) are only MP. The product heed not be transported, so line BQ can be ignored.

Figure 6.2(*c*) suggests that, under these conditions of tapering transport costs, industries will tend to locate mostly at the sources of their raw materials or at the markets for their products, and not at some point in between. The tapering cost curves give a humped total transport cost curve. Also one set of terminal costs is saved at each of locations R and M.

Figure 6.2(*d*) illustrates a case in which it is more costly to transport the manufactured product than the raw material, which very frequently is the case in real life. Total transport costs are least if the factory is located at the market (M) where the only transport costs are those involving the raw material (AP) and only one set of terminal costs (RA) is incurred. In this case the location with the highest transport costs has moved away from X towards the source of the raw material.

In some cases the transport route between the raw material source and the market for the product may include a transhipment point where it is necessary to change from one mode of transport to another. This is often a port where land transport meets water transport. Transhipment may also occur where road transport meets rail transport or air transport.

Figure 6.3(*a*) shows the cost of transporting a raw material from its source (R) to a market (M) through a transhipment point (T) where costs are incurred as the material is transferred from one type of transport to another. Figure 6.3(*b*) shows similar costs for the transport of a manufactured product to the market. This again may have to be transhipped at T. Transport of either raw material or product through the transhipment point involves two separate journeys; thus some of the benefit of tapering transport costs is lost. When the journey continues beyond T the steeply rising part of the transport cost curve begins again.

In this case it is possible for total transport costs to be quite low if the factory is located at the transhipment point (T). Here there will be no transhipment costs since neither the raw material nor the product will pass straight through it. There will be terminal and line-haul costs for bringing the raw material to T where it is manufactured; then there will be terminal and line-haul costs for transporting the product to the market. Figure 6.3(*c*) shows that it is possible for total transport costs at the transhipment point to be as low as at either the raw material source or the market.

If the raw material is particularly cheap and

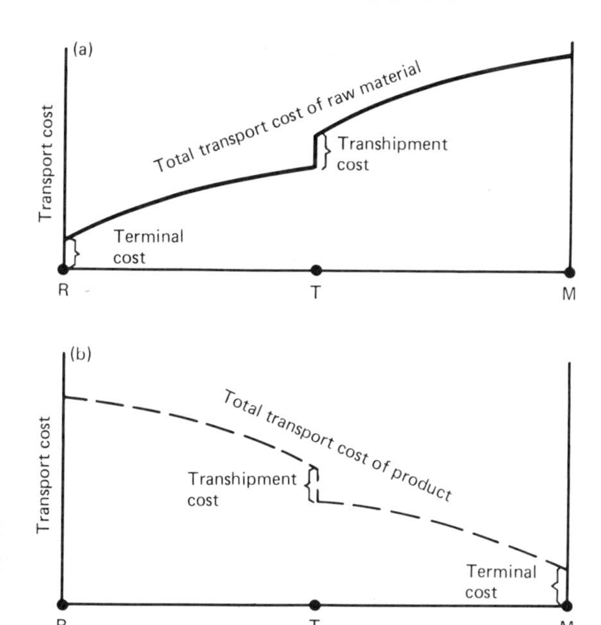

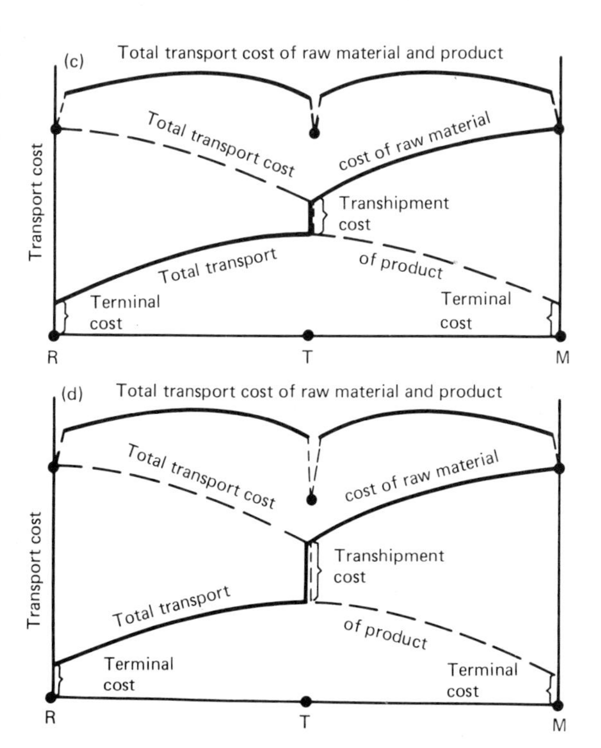

Fig. 6.3 The influence of a transhipment point upon industrial location

Transport oriented industry on the banks of the Manchester Ship Canal

bulky and therefore relatively expensive to tranship, it is possible for total transport costs at the transhipment point (T) to be lower than at either the raw material source or the market. This case is illustrated in Figure 6.3(*d*).

This principle helps to explain why so many industries processing bulky raw materials are located at seaports or lake or river ports. Examples are flour milling, the iron and steel industry and oil refining. However, technological developments in the bulk handling of materials at transhipment points may reduce transhipment costs and therefore reduce the advantage of the transhipment point as a location for manufacturing industry. For example, the modern use of containers may in some cases encourage manufacturing industry to locate either at the raw material source or at the market.

THE INFLUENCE OF PROFITS

Just as costs of production may vary from place to place so also may a firm's revenue from the sales of its product. Variations from place to place of a firm's production costs can be represented by a space-cost curve (Fig. 6.4); variations in a firm's revenue can be represented by a space-revenue curve (Fig. 6.4). In some locations the firm's revenue may be greater than its costs and it may make a profit. In other locations its costs may be greater than its revenue and it may make a loss and eventually have to close down.

Figure 6.4(*a*) shows a case in which a firm's

revenue is constant over the area shown, but its costs vary. Maximum profit is earned where production costs are least at location X. However, under these conditions it is not always necessary for the firm to locate at its least cost location. Figure 6.4(*b*) shows a case in which costs are constant over space. The maximum profit location in this case is at the maximum revenue location (X).

It is possible for the firm to make the maximum possible profit without locating at either the maximum revenue or the least cost location. In Figure 6.4(*c*) maximum profit (greatest difference between revenue and costs) is earned at location X, whereas the least cost location is Y and the maximum revenue location is Z.

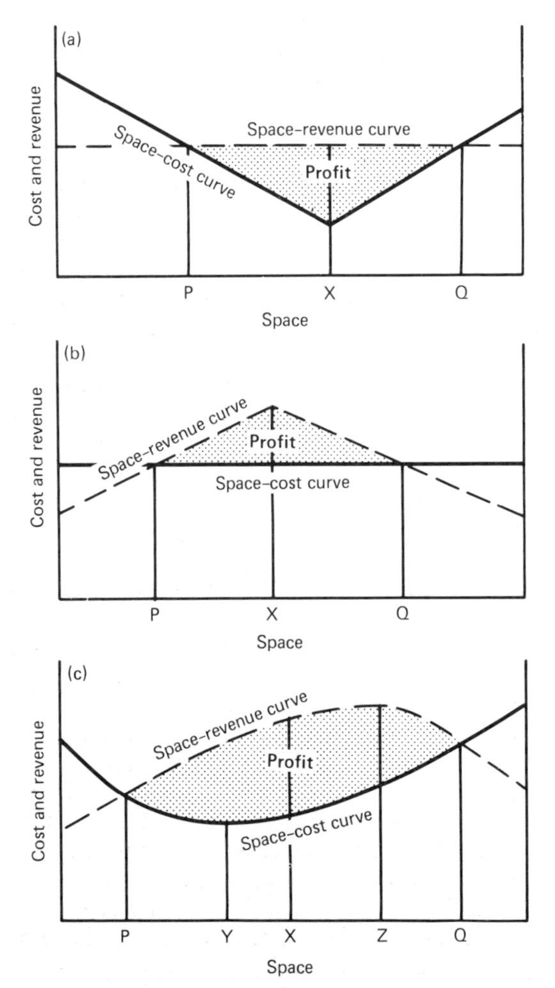

Fig. 6.4 The influence of profits upon industrial location

In each of these cases a firm may survive without locating at its least cost location or its maximum profit location. A location at Y or Z in Figure 6.4(*c*) for example would be quite satisfactory even though it is not the best possible location. Anywhere between P and Q in all three cases in Figure 6.4(*c*) would be satisfactory.

It is unlikely that very many firms in fact seek out the maximum profit locations for their plants. They are satisfied with a reasonable profit level. They may allow their location decisions to be influenced by other factors within the general area within which it is possible to make profits.

THE INFLUENCE OF A LABOUR SUPPLY

The cost of labour forms an important part of total production costs in certain industries in which it has been difficult to introduce mechanization. These include the textile and clothing industries, the manufacture of shoes, and shipbuilding and ship repairing. Such industries' locations may be influenced by the existence of an efficient, low-cost labour supply. They also commonly require a labour supply with particular skills. Other examples are the cutlery, furniture and watch-making industries.

A labour supply is much less important for industries such as oil refining and the manufacture of chemicals. In these cases capital (machines of various kinds) is able to be used instead of labour. In general, as automation and mass production in manufacturing industry have developed a labour supply has become less important as a factor influencing the location of industry.

In general, people are not very willing to change their jobs or to move home from one part of the country to another. Labour is said to be 'immobile'. Hence, an industry that needs a particular type of labour must locate in the area where this labour exists, or at least within commuting distance. Large cities are therefore particularly attractive to many industries. Here they will find a wide variety of labour skills and they will also have a better chance of attracting able managers and technical staff.

Labour costs often vary from place to place. This is not necessarily a reflection of differences in wage levels, but may be related to general attitudes to work, levels of absenteeism and differences in productivity. A variety of labour-intensive industries have developed rapidly in countries of the Far East such as Hong Kong, Taiwan and South Korea, helped by their low labour costs.

Some industries such as textiles have traditionally been attracted to areas where a large supply of female labour was available. These have often been areas dominated by industries employing largely men, such as mining or shipbuilding areas.

THE INFLUENCE OF A SUPPLY OF CAPITAL

There are two main kinds of capital available to manufacturing industry:

(*a*) Money capital can be used by a firm to purchase new buildings and equipment. This money may be made available by the firm either ploughing back its profits or borrowing in various ways from financial institutions including the banks.
(*b*) Physical capital consists of buildings and equipment which have been created in the past, and which already exist in a particular place.

Money capital is fairly mobile. If a firm is intending to set up a new factory it is usually reasonably free to choose where the factory shall be located, without regard to the place where the money capital originated. If the firm borrowed money in London this does not mean that the factory must be built in London. Very large, powerful firms in particular are free to choose factory locations regardless of the origin of their capital.

In some cases, however, money capital may not be so mobile. If a government provides grants or loans for the building of new factories it may stipulate that the factories should be built in particular areas of the country according to its regional policy. Also, money capital is often not very mobile internationally. It may be difficult for a developing country to obtain capital from overseas to develop its own manufacturing industries.

Physical capital is not at all mobile. If an industry wishes to use this capital equipment it can only do so by locating where the physical capital exists. Thus, the existence of factories and other items of fixed capital in particular places tends to ensure that manufacturing industries continue to exist in these areas. Factories originally built in the nineteenth century for the textile industry may be a most important factor in attracting new industries to the area. In Bolton, many cotton spinning mills and bleachworks have become vacant as the textile industry has contracted, and these premises have attracted a wide variety of modern industries. Similarly, it is very difficult for an industry such as the steel industry to change its location. Where steel works already exist a great deal of physical capital is tied up, not only in the steel works itself, but also in the housing and general infrastructure which has been built up in the area. Such problems have existed in the attempts which have been made to reorganize the British steel industry. Physical capital therefore tends to perpetuate the existing spatial pattern of manufacturing industry. It encourages geographical inertia.

THE INFLUENCE OF ECONOMIES OF AGGLOMERATION

In many cases, as industries develop in particular locations in response to the various factors outlined above, their very existence makes the locality more favourable for the growth of new industries. Many industries benefit greatly from nearness to other industries. These benefits are referred to as *agglomeration economies* (the savings made through firms collaborating with one another within an industrial region). These economies derive to a great extent from the *linkages* which exist between different industries. There are many kinds of industrial linkages, as indicated below.

Some firms use the products of other firms as their raw materials. Thus, in an industrial region, one industry has a supply of raw materials close by and the other industry has a market for its products close by. So both of these industries benefit. Linkages between textile industries and clothing industries, and between the iron and steel industry and the motor car industry, are of this type. A great variety of different industries use the products of the iron and steel, oil refining and chemical industries as their raw materials.

In some cases, a number of different industries may produce a variety of industrial products which are combined together to produce a highly complex finished product. The motor car industry is related in this way to industries manufacturing steel, rubber and plastic products, electrical equipment and glass products, to name only a few.

Hence, because of these linkages, and many others, a great variety of firms derive benefit from a location in an industrial region in which many different industries exist. In such an industrial region industries may obtain the benefits of *external economies.* These are all the general advantages, not specifically industrial, which exist in a large industrial region. In such a region an industry may be able to draw upon a large labour supply with many varied skills. Also, educational institutions such as universities, and research organizations may exist in the area to provide valuable services to industry. Industrial regions are usually supplied with efficient transport services between their various districts and also with other industrial regions and the capital city. The motorway system of England, for example, links together the great industrial conurbations most efficiently. No industrial region is more than two hours' travel time from another such region or a major port.

The great advantages of agglomeration within such industrial regions means that major changes in the spatial distribution of industries is very unlikely. Firms cannot afford to move away from the location of these agglomeration economies and external economies. Thus agglomeration is another factor which tends to encourage industrial inertia.

THE INFLUENCE OF GOVERNMENT POLICIES

Regional differences in the level of industrial development exist in all countries. Hence there are differences in unemployment rates, income levels, emigration rates, etc. Governments generally feel that it is undesirable for there to be great contrasts in living standards between the regions of a country. They therefore adopt 'regional policies' to encourage economic development in the problem areas and to discourage congestion in the richer areas. These policies are described in Chapter 3 on page 55. There are two main types of problem areas in most developed countries. Some have always been underdeveloped; others have problems because their industrial structure has failed to adapt to modern conditions.

Areas such as the Scottish Highlands and central Wales in Great Britain and the Mezzogiorno in southern Italy have failed to keep pace in terms of economic development with other parts of their respective countries. They have never had a significant degree of industrial development, being remote from the major centre of national life and lacking adequate transport facilities. Hence they have suffered high rates of emigration. Their respective governments have set up organizations intended to promote economic development, including industrial development, in these areas. Thus, new industry has moved into the Moray Firth area, including an aluminium works at Invergordon, and iron and steel and petrochemical industries have been established in southern Italy. Also, programmes of road building have been carried out, partly with the intention of making these areas more attractive to industrialists.

Other areas formerly had thriving industries in the nineteenth century, such as the manufacture of textiles, coal mining, iron and steel and shipbuilding. They frequently possess large areas of outdated houses and factories and a good deal of derelict land. Many of them are unattractive to industrialists considering a new site for a modern industry. Such areas however do possess resources that are useful to industry. They commonly have a large labour supply with varying skills and reasonably adequate transport systems. In the interests of using the country's labour and capital

Industry occupying cramped sites near an old town centre

resources as fully as possible, governments try to encourage industrial firms to set up in such areas and they offer firms financial incentives (grants and low-interest loans) to establish factories there. The British government has carried out such a policy since the 1930s when there was serious unemployment in the old coal-based industrial regions. At the same time, it has often discouraged further industrial development in the Birmingham and London areas.

Perhaps the most outstanding examples of the influence of government on industrial location are found in the centrally planned economies such as the USSR and other Warsaw Pact countries. Here the central planning authority has determined the location and the degree of development of various industries in different parts of the national territory.

6.2 The Weber model of the location of industry

Weber was a German economist who was born in the latter part of the nineteenth century. His ideas on industrial location were first published in 1909.

WEBER'S ASSUMPTIONS

Like all other model builders, Weber expressed his ideas in relation to a simplified version of the real world. In order to achieve this he set out certain assumptions which were intended to simplify the complexity of reality.

Like von Thünen, Weber assumed the existence of a perfectly uniform land surface across which it was equally easy to move in any direction. Movement need not take place along the channels of a transport network but could take place in any direction at all.

He also assumed that transport costs were a simple function of weight of materials carried and distance. Transport costs are expressed as tonne kilometres. They are directly proportional to weight and distance.

Certain raw materials exist on this uniform plain. Some of these raw materials are 'ubiquitous' and are available everywhere. Others are 'localized' and only occur at particular locations.

Some raw materials are 'pure' which means that, on being manufactured, the whole of their weight is included in the finished product. Others are 'gross' which means that only part of their weight is included in the finished product. Thus, during the manufacture of gross raw materials, some waste material of no value is produced.

Markets for the industry's manufactured products exist at particular *points* on the uniform plain and do not extend over an *area* of the plain.

Entrepreneurs try to locate so as to minimize their total transport costs. They have full knowledge of the transport costs and of the location and characteristics of the raw materials and the market.

EXAMPLES OF LEAST COST LOCATIONS

The following examples are cases in which the industry uses two raw materials (A and B) and its product is sold at a single market.

If both raw materials are ubiquitous

If both raw materials are ubiquitous (Fig. 6.5(*a*)) the least cost location is always the *market*, whether the raw materials are gross or pure. Since both raw materials are ubiquitous, they must both

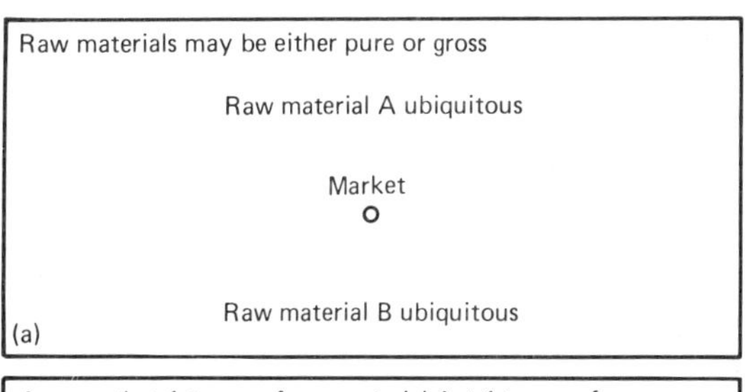

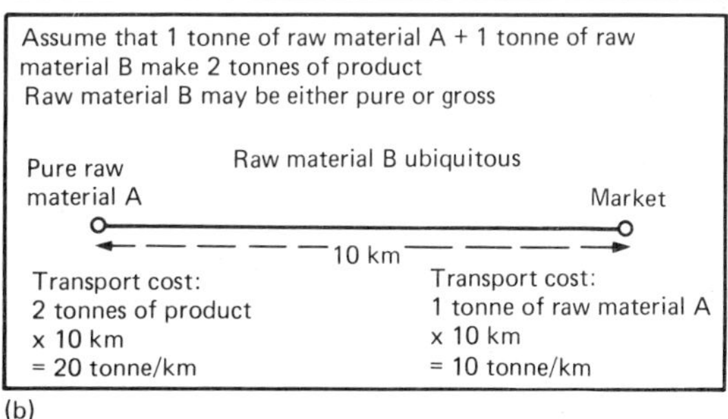

Fig. 6.5 Least cost location if at least one raw material is ubiquitous

exist at the market. Since the product is also sold at the market, there are no transport costs at all.

If one raw material is localized and the other ubiquitous

The case where one raw material is localised and the other ubiquitous is illustrated in Figure 6.5(*b*).

If the localized raw material (A in Fig. 6.5(*b*)) is pure, the least cost location is the *market*. At the market only the pure raw material A needs to be transported. If the industry were to be located at raw material A the product which has to be transported from here to the market must have a greater weight than the pure raw material A.

If the localized raw material (A in Fig. 6.5(*b*)) is gross, the least cost location may be either at the market or at raw material A. If the weight of the product (transported to the market) is greater than the weight used of the gross localized raw material A, the least cost location is the *market*. If the weight of the product is less than the weight of localized raw material A, the least cost location is at *the source of raw material A*.

In relation to such cases as these, Weber devised a Material Index, calculated by dividing the weight of localized raw materials used in the industry by the weight of the product. A Material Index greater than 1 indicates a tendency towards a raw material location. A Material Index less than 1 indicates a tendency towards a market location.

If both raw materials are localized and pure

If both raw materials are localized and pure the material index must always have a value of 1.

If the raw materials and the market are arranged in a straight line the least cost location must be the *market*, provided that it is situated anywhere between the two localized materials.

If the raw materials and the market are arranged in a straight line and both raw materials are located on the same side of the market (as in Fig. 6.6(*a*)), the *market* is still a least cost location, but several other locations have the same transport costs as the market.

If the raw materials and the market are not arranged in a straight line but instead form the apices of a triangle, the least cost location is always the *market*. This case is illustrated in Figure 6.6(*b*).

If one raw material is pure and the other gross and both are localized

If one raw material is pure, the other gross and both are localized, the industry tends to be attracted towards the source of the gross raw material.

Figure 6.7(*a*) shows one example of this situation in which raw material A is pure and raw material B loses 50% its weight during processing. Of the locations shown the *intermediate location* has the lowest transport costs. In fact, the least cost location of all is situated within the triangle, a short distance from the intermediate location in the direction of raw material A.

Figure 6.7(*b*) shows a case in which raw material B loses much more of its weight during processing. In this case the least cost location is clearly at the source of *raw material B*. A simple rule is that the least cost location is at raw material B if the weight used of raw material B is equal to or greater than the combined weights of raw material A and the product.

If both raw materials are localized and gross

If both raw materials are localized and gross the industry tends to be attracted towards the raw material that loses the greater proportion of its weight during processing. There may be a case in which the weight used of one raw material is equal to or greater than the combined weights of the other raw material and the product. In this case the least cost location is at the source of the raw material of which the greater weight is used.

THE DISTRIBUTION OF TRANSPORT COSTS OVER SPACE

It is also possible to show how the transport costs of raw materials and product vary from place to place by drawing a kind of 'contour' map in which the 'contour lines' represent transport costs and not altitude.

One pure localized raw material and a single market

The case of one pure localized raw material and a single market is illustrated in Figure 6.8. One tonne of raw material is manufactured into one tonne of product which is then delivered to the market.

Figure 6.8(*a*) shows the transport costs (in tonne kilometres) for one tonne of the raw material. These are shown as concentric circles centred on the source of the raw material (RM). Thus, the cost of transporting one tonne of the raw material

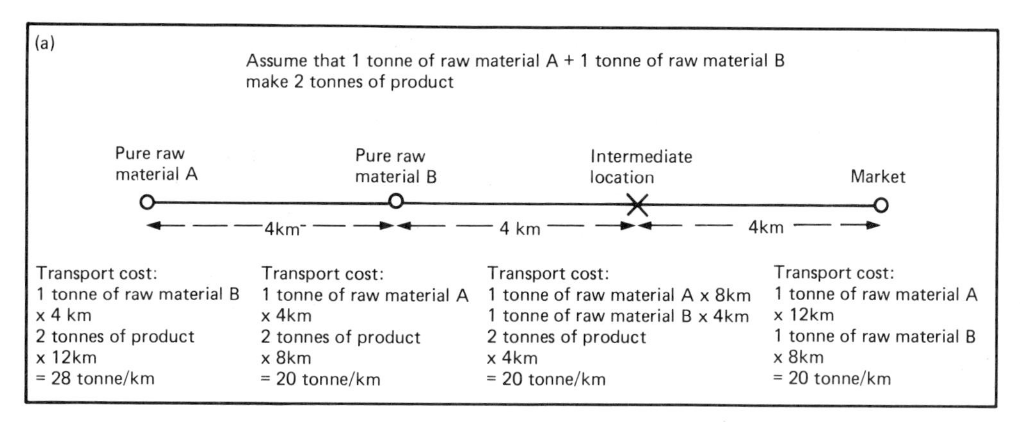

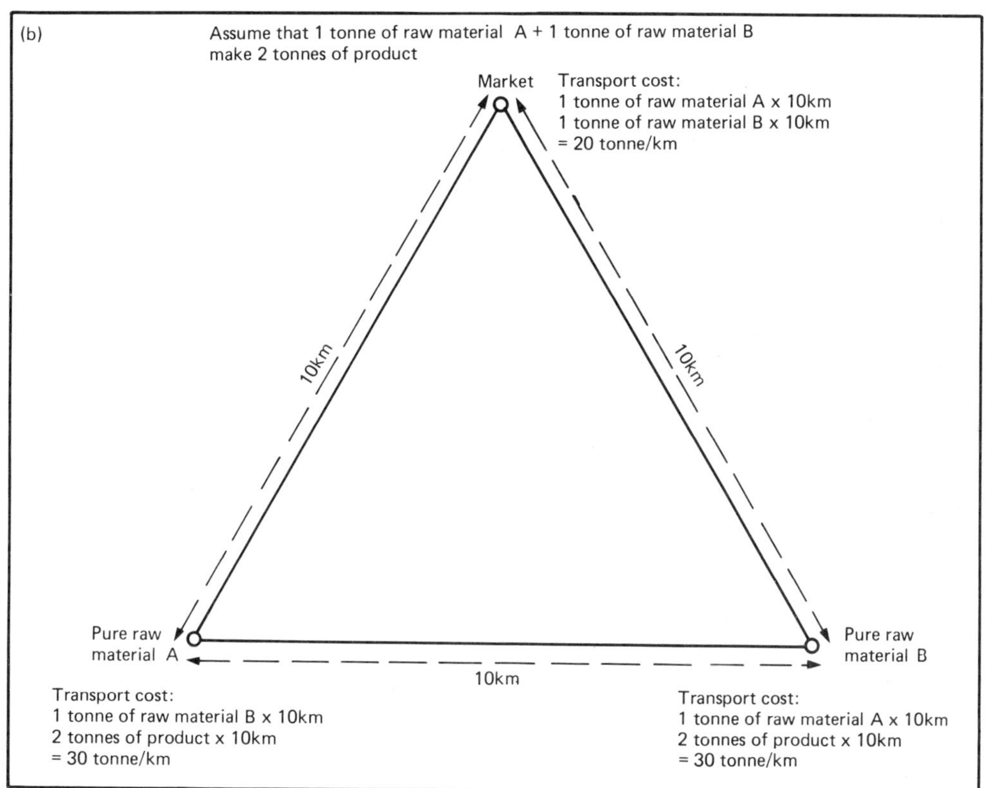

Fig. 6.6 Least cost location if both raw materials are localized and pure

to the market is 5 tonne kilometres. These concentric circles are known as *isotims*.

Figure 6.8(*b*) shows the transport costs for one tonne of the product from any part of the area to the market (again measured in tonne kilometres). Thus, the cost of transporting one tonne of product from the source of the raw material to the market is 5 tonne kilometres.

It is clear from Figures 6.8(*a*) and 6.8(*b*) that the transport costs of the raw material and the product are equal because the isotims are all the same distance apart.

Figure 6.8(*c*) is the result of combining Figure 6.8(*a*) with Figure 6.8(*b*). Thus, it has become possible to insert a new set of lines showing total transport costs (of both raw material and product)

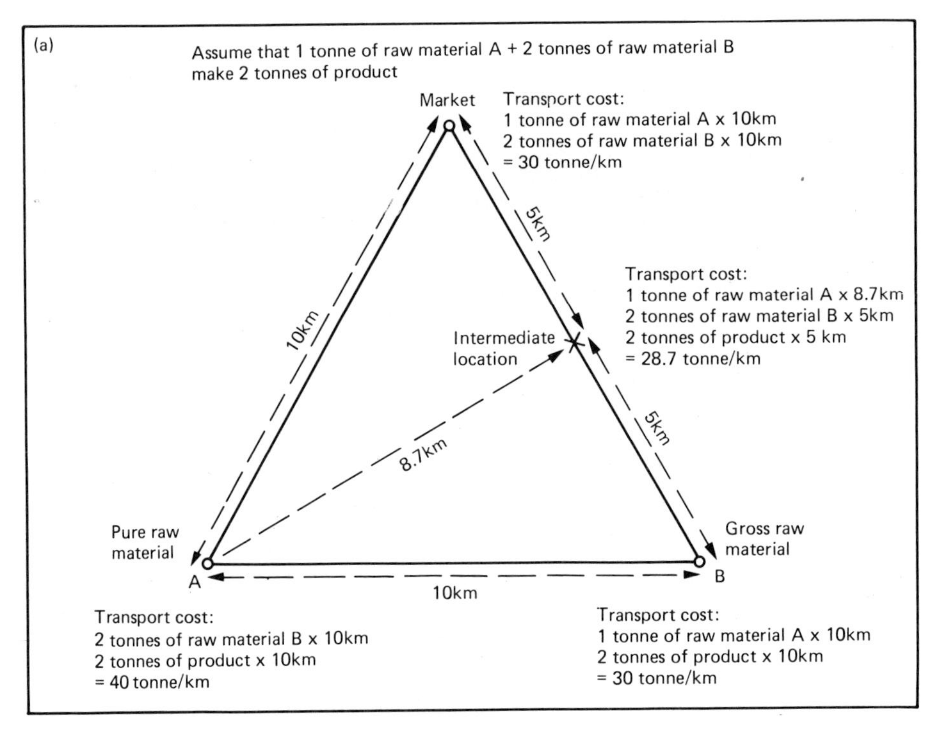

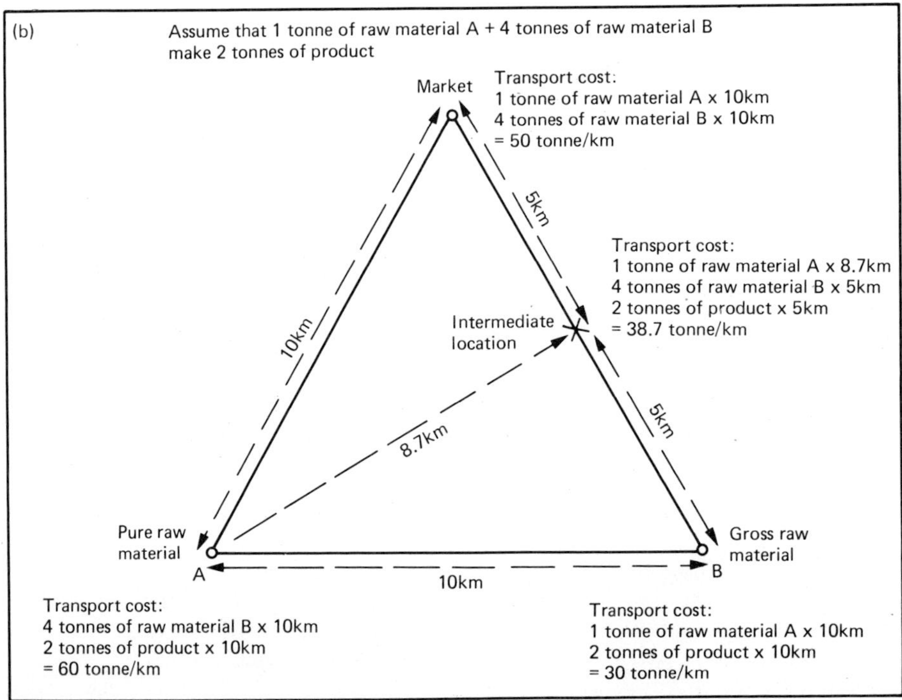

Fig. 6.7 Least cost location if one raw material is pure, the other gross and both are localized

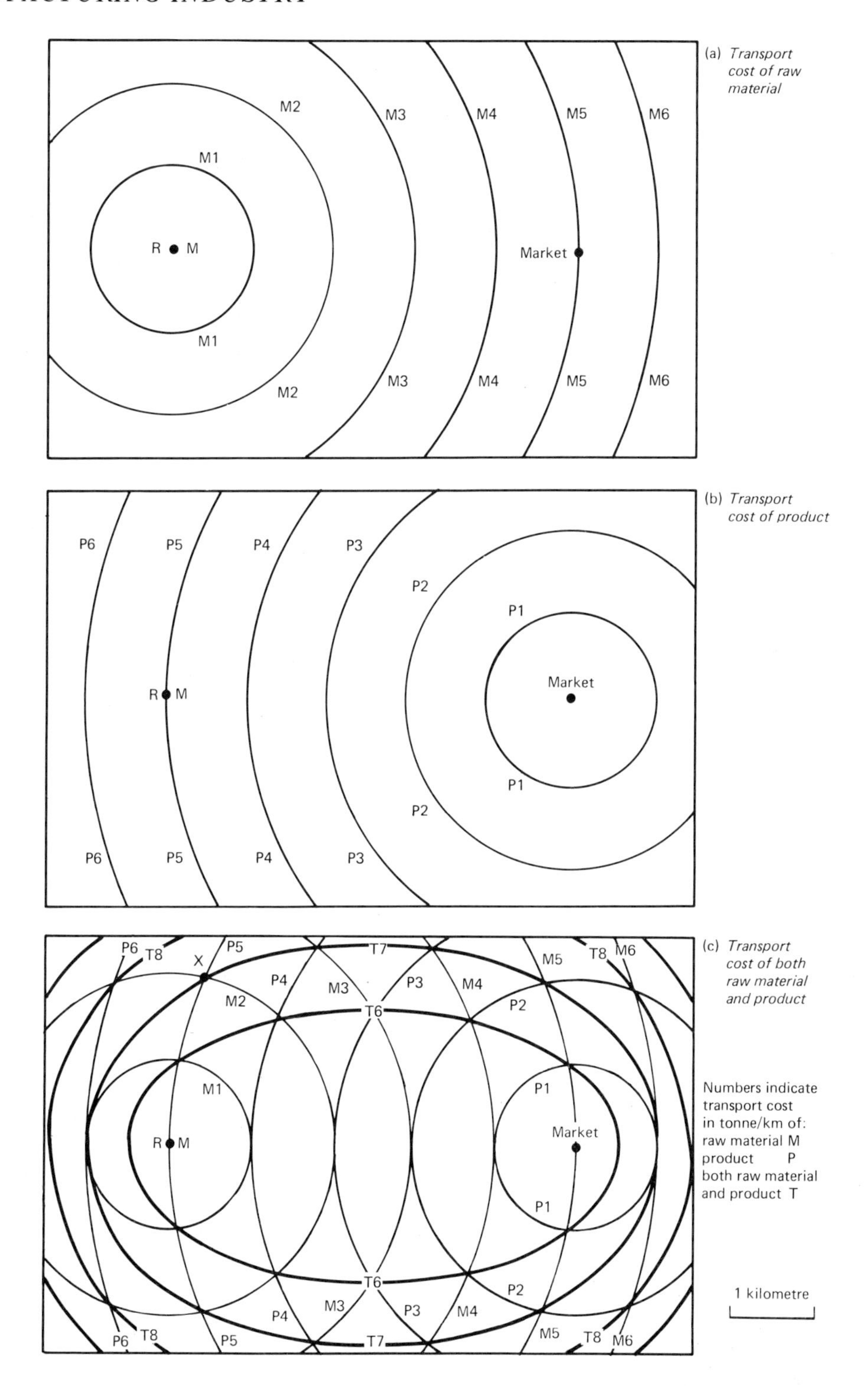

Fig. 6.8 Isotims and isodapanes. Equal transport costs for raw material and product

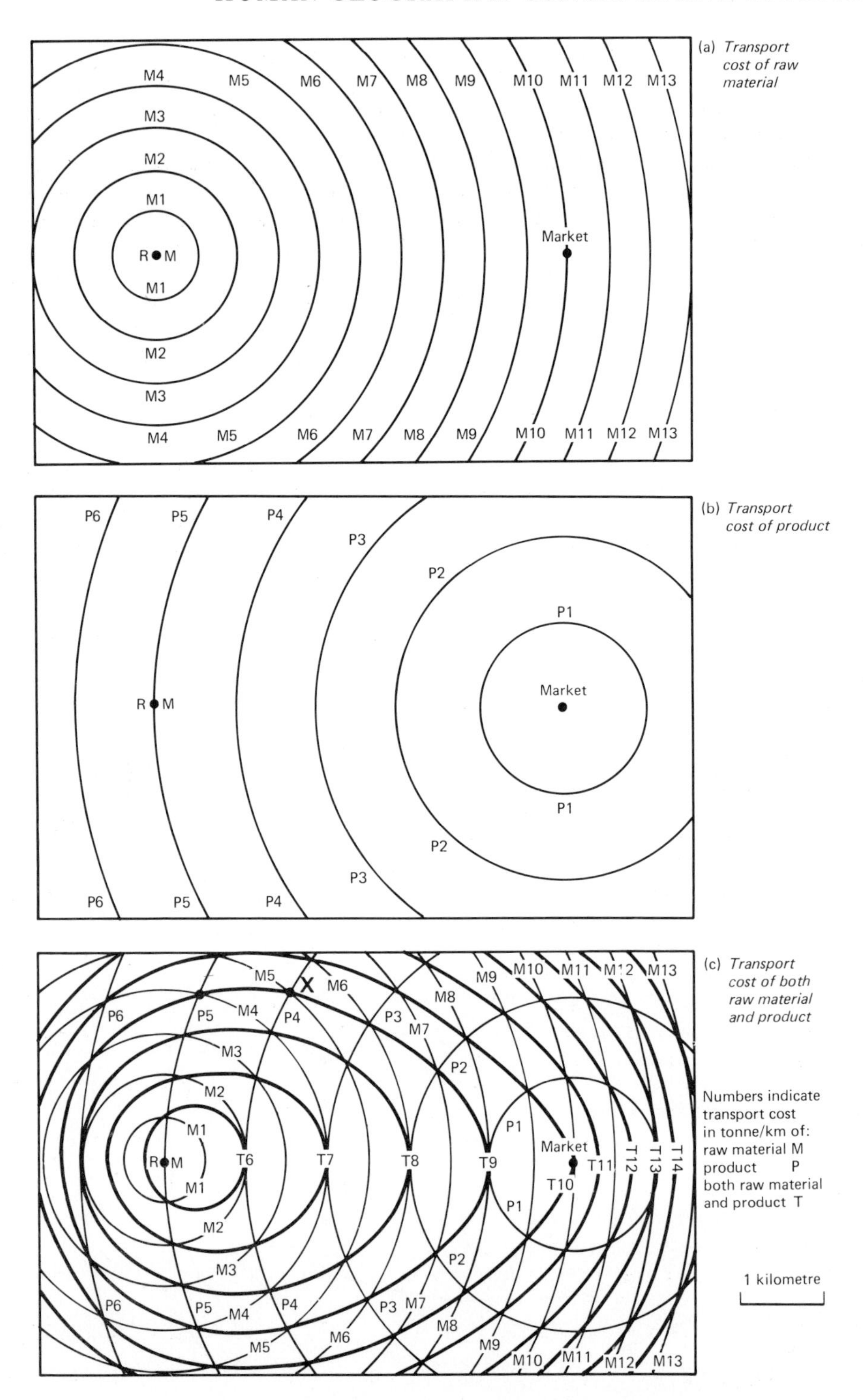

Fig. 6.9 Isotims and isodapanes. Unequal transport costs for raw material and product

in all parts of the area. These lines of total transport costs are known as *isodapanes*. If an industry were to be located at the position marked X in Figure 6.8(*c*) its total transport costs per tonne of product would be 7 tonne kilometres (2 tonne kilometres (M2) for the raw material and 5 tonne kilometres (P5) for the product).

In Figure 6.8(*c*) it can be seen that the lowest possible total transport costs are 5 tonne kilometres at both the raw material source and the market, and also at all points along a straight line from one to the other.

One gross localized raw material and a single market

In Figure 6.9 *two* tonnes of raw material are manufactured into one tonne of product which is then delivered to the market.

The transport costs for two tonnes of the raw material are shown in Figure 6.9(*a*). Compared with Figure 6.8(*a*) the isotims are much closer together, showing that transport costs rise more steeply with distance. This is because two tonnes of raw material have to be transported instead of one tonne. Thus, it costs 10 tonne kilometres to transport two tonnes of the raw material the five kilometres to the market.

Figure 6.9(*b*) shows the transport costs for one tonne of the product. These are exactly the same as in Figure 6.8(*b*) because only one tonne of product is transported.

Figure 6.9(*c*) consists of Figure 6.9(*a*) combined with Figure 6.9(*b*), and the isodapanes showing total transport costs have been added.

Total transport costs are least at the source of the raw material (RM) where they are 5 tonne kilometres (P5). They increase steadily to 10 tonne kilometres (M10) at the market. Total transport costs also increase in all other directions away from the source of the raw material.

In this case it is clearly an advantage for the industry to be located relatively near the raw material. At the position X, for example, total transport costs (9 tonne kilometres) are still smaller than at the market even though a much greater travel distance is involved.

The influence of labour costs and agglomeration economies

In addition to the influence of transport costs on industrial location, Weber also considered the possible influence of varying labour costs and economies resulting from agglomeration (page 113). He considered that savings made in these respects could cause an industry not to locate at its least transport cost location.

In Figure 6.9(*c*) for example the transport cost at the location marked X is 9 tonne kilometres, whereas at the raw material source (the lowest transport cost location) it is only 5 tonne kilometres. Hence, if labour or agglomeration economies at location X are able to save more than the equivalent of 4 tonne kilometres (9 minus 5) the industry will have lower costs at X than at the least transport cost location.

If the savings resulting from labour or agglomeration economies amounted to exactly 4 tonne kilometres, the industry could reduce its total costs by moving away from the raw material source (Fig. 6.9(*c*)) to any place where these economies were available provided that this place was located within the 9 tonne kilometre (T9) isodapane. In

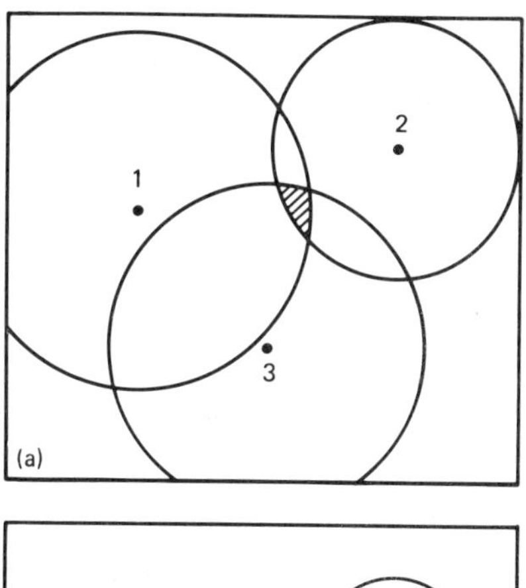

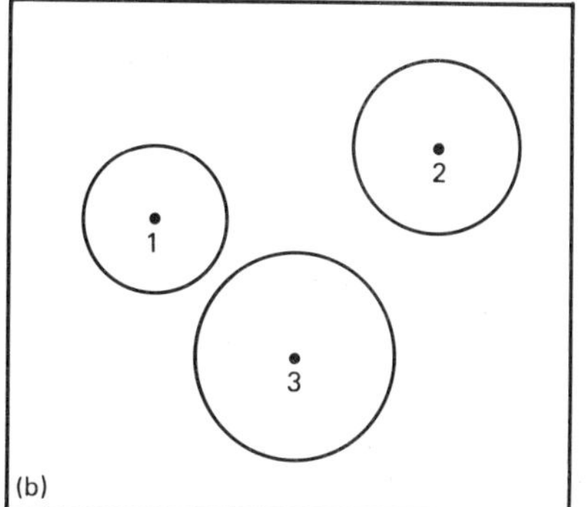

Locations 1, 2 and 3 are least transport cost locations.
Circles are critical isodapanes.
Shaded area shows where agglomeration will reduce costs in all cases.

Fig. 6.10 Critical isodapanes and agglomeration

this case, this isodapane would be referred to as the *critical isodapane*. If the savings amounted to only 3 tonne kilometres, the 8 tonne kilometres (5 + 3) would become the critical isodapane.

A number of different industries may agglomerate at one location (not their least cost location) if their critical isodapanes intersect, as shown in Figure 6.10(*a*). None of these industries are at their least transport cost location. In Figure 6.10(*b*), agglomeration economies are not sufficient to offset the increases in transport costs, so the industries remain at their least transport cost locations and do not agglomerate. In this case of course their critical isodapanes do not intersect.

Exercises

1. Discuss the comparative importance of raw materials, power supplies, markets and labour supply as factors influencing the location of the following industries:

(*a*) the manufacture of cement;
(*b*) shipbuilding;
(*c*) the manufacture of scientific instruments;
(*d*) the rust-proofing treatment of new cars.

2. Explain why it is rare for an industry to be ideally located in relation to its raw materials, its power supplies or its markets. Illustrate your answer by discussing actual examples.

3. (*a*) Explain why it is common for various industries to cluster together to form industrial regions.
(*b*) Name some industries that are only rarely found in industrial regions and explain why.

4. Discuss the relevance of the Weber model to the understanding of the present-day distribution of manufacturing industry.

5. In the light of the Weber model describe and explain the conditions under which an industry's least-cost location may be

(*a*) at the market for the product;
(*b*) at the location of a gross raw material;
(*c*) at the location of a pure raw material;
(*d*) at a location intermediate between raw materials and market.

6. To answer this question refer to Figure 6.11 and apply the principles of the Weber model. Assume that there is a single raw material location and a single market for the product.

(*a*) If the raw material is pure
   (i) state the transport cost (in tonne kilometres) per tonne of product delivered at the market for an industry located at each of points A to D.
   (ii) describe the location(s) at which transport costs are minimized, and explain your reasoning.
(*b*) If the raw material is gross and loses 50 % of its weight during manufacture
   (i) state the transport cost (in tonne kilometres) per tonne of product delivered at the market for an industry located at each of points A to D.
   (ii) describe the location at which transport costs are minimized, and explain your reasoning.

7. Figure 6.12 shows the location of a single raw material source, a single market and a transhipment point. The raw material is pure.

(*a*) Transport costs for both raw material and product are £1 per tonne per 10 kilometres. Transhipment costs are given in Figure 6.12. Calculate the transport cost per tonne of product delivered at the market for an industry located
   (i) at the source of the raw material;
   (ii) at A;
   (iii) at T;
   (iv) at the market.
   Describe the spatial variations in transport costs over the area shown in Figure 6.12.
(*b*) Transport costs for the raw material rise to £5 per tonne per 10 kilometres, but the cost of transporting the product remains at £1 per tonne per 10 kilometres. Describe and explain the spatial variations in total transport costs over the area shown in Figure 6.12.

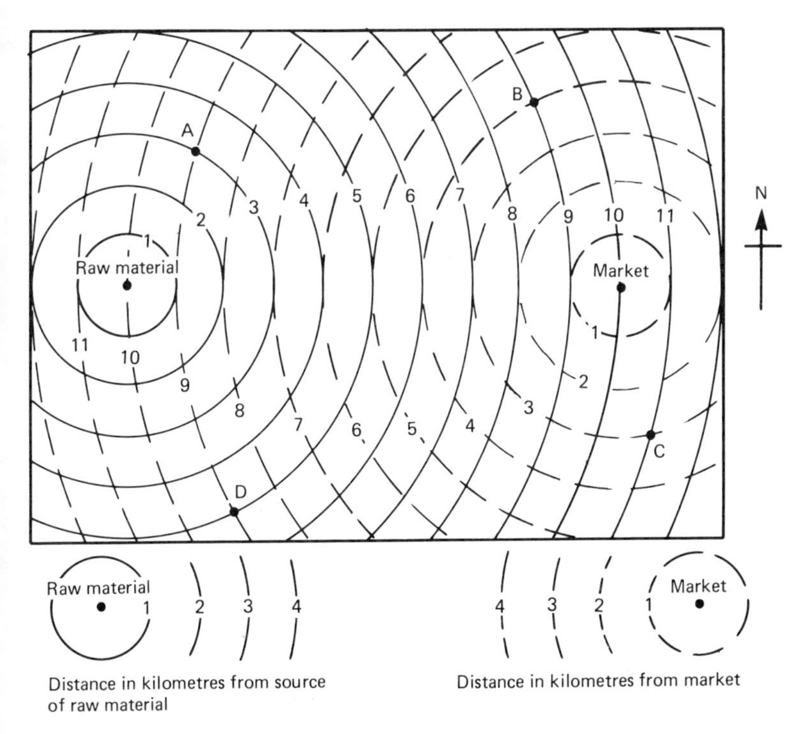

Fig. 6.11 Location of a raw material source and a market

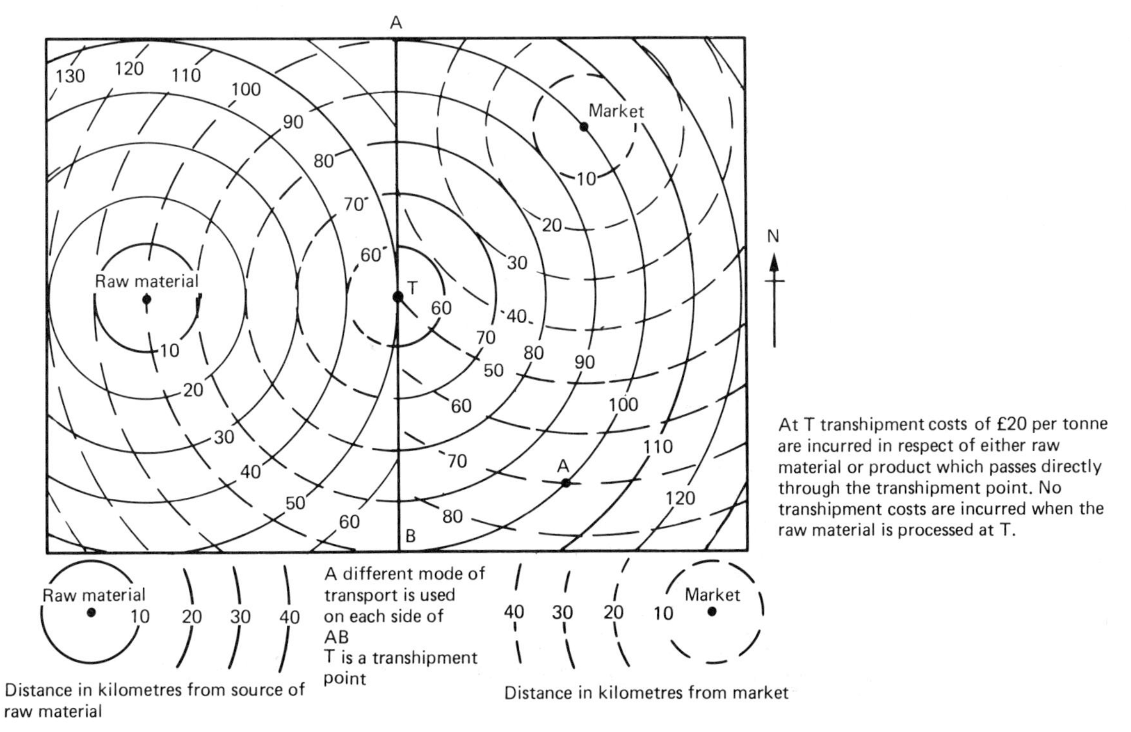

Fig. 6.12 Location of a raw material source, a market and a transhipment point

7 International trade and transport

7.1 International trade

WHY INTERNATIONAL TRADE EXISTS

International trade is a type of specialization or 'division of labour'. In ordinary life we take it for granted that a person spends the greater part of his life working in a particular occupation and, with the money he earns, he buys his various needs from people who are working in other occupations. A teacher, for example, is not expected to grow his own food, make his own clothes or repair the roads that he uses. Similarly the people who grow food, make clothes or repair roads are not expected to educate their own children. Thus, people may become skilful and productive in particular occupations, and the whole community will benefit from this.

Similarly, it is to the general benefit of the world community if different countries specialize in the production of certain commodities or the provision of various services. Thus Switzerland specializes in the manufacture of watches, and Spain, with its sunny climate, in the provision of holidays. Switzerland exports its watches to the rest of the world and Spain sells (exports) holidays to people from all parts of Western Europe. Without international trade each country would have to be self-sufficient and living standards would therefore be lower. Great Britain, for example, could never compete climatically with Spain for holidays.

Many different kinds of specialization can take place, each one giving rise to trade.

Some natural resources, particularly minerals, are concentrated in certain countries. Examples are tin (in Malaysia, Bolivia and Thailand) and bauxite (in Australia, Guinea and Jamaica). Hence these countries supply the rest of the world, by trade, with these commodities. The OPEC countries have a virtual monopoly of the world's petroleum supplies.

Some crops can only be grown economically in certain types of climates. South-east Asia produces over 90% of the world's natural rubber, West Africa over 60% of the world's cocoa beans, the Latin American tropical countries over half of the world's bananas. Although it is quite possible to grow bananas in heated greenhouses in Britain, it is much more efficient to import them from Central America and pay for them by exporting manufactured goods.

Countries also differ in the amounts of capital and land that they possess. Advanced countries have accumulated great capital resources, so they can produce manufactured products which are needed by less developed countries. Very large advanced countries, such as Canada, have large amounts of land as well as capital. So, by extensive farming (page 59) they can produce enormous amounts of farm produce, such as wheat, which they export to smaller, more densely populated countries where farming is more intensive and concentrates on dairying and market gardening.

THE DIRECTION OF INTERNATIONAL TRADE

General evolution

Large-scale international trade only began in the nineteenth century. At this time the Industrial Revolution was taking place in Europe, and various European countries were establishing colonies in various parts of the world. Much of the world's trade consisted of the import of food and raw materials from these colonies and the export of manufactured goods in return.

Also, enormous areas of newly developed land were opened up in North and South America and Australia, and these became suppliers of food and raw materials to Europe. In addition the United States and Japan developed industrially.

Great changes have taken place since the middle of the twentieth century. The traditional trade between European countries and their colonies has suffered a relative decline in importance. The old empires of the European countries have disappeared. Former colonies have usually become rather small, economically weak members of the Third World. Trade between the advanced industrial countries of the world has greatly increased and has replaced the former 'colonial' pattern of trade. The advanced market (non-communist) economies are now responsible for two-thirds of all world trade. The developing market (non-communist) economies are responsible for less than a quarter. The share of centrally planned (communist) economies is about 10 %. The main recent trend has been the great increase in the value of exports of petroleum particularly to advanced market economies from the OPEC countries (page 96).

Since the 1940s there has been a general movement towards removing restrictions to trade such as the payment of tariffs in respect of imports entering a country. In 1948 the General Agreement on Tariffs and Trade (GATT) came into operation. Its purpose was to increase the freedom of trade by reducing tariffs and generally removing barriers to trade between countries. Also, groups of countries have joined together to form 'common markets' within which trade can flow more freely between member countries (page 31). These include the European Economic Community (EEC), the European Free Trade Association (EFTA) and the Latin American Free Trade Association (LAFTA).

The pattern of international trade

Figure 7.1 shows the broad pattern of trade between the advanced market economies, the developing market economies (including the OPEC countries) and the centrally planned economies. Almost half of total world trade flows between the member countries (largely industrialized) of the advanced market economies. About one-third of world trade flows between the advanced market economies and the developing market economies, the greatest elements here being petroleum exports from the OPEC countries (9.7 %) and exports largely of manufactured goods from the advanced to the developing market economies (10.4 %). The remaining flows are much smaller. Trade between the advanced market economies and the centrally planned economies amounts to only about 6 % of the world total. Very little trade takes place between the centrally planned economies and the developing market economies.

Figure 7.2 shows the pattern of world trade in rather more detail. Here it is clear that the EEC is the focal point of international trade with high-volume links with all the other major trading units in the world. Altogether the trade of the EEC makes up about one-third of the world's total. As can be seen in the diagram, much of this is trade between the various member countries of the EEC. This is partly due to the fact that the EEC is made up of very small countries and trade between them is classed as 'international' as soon as it crosses one of their national boundaries. In the USA, for example, which is six times greater in area than the EEC, trade between its different regions is not classed as 'international'. Nevertheless, the EEC is the only unit that has relatively high-volume trade links with all the other major trading units of the world. Its largest flows of imports come from the OPEC countries, EFTA, which is also in Europe, and the USA. Its major flows of exports also go to EFTA, OPEC and the USA.

The USA forms a secondary focal point of world trade. It has strong trading links with Canada and the EEC and slightly weaker ones with the OPEC countries and Japan. Japan's trade is mostly with the OPEC countries (mainly for imports) and the USA.

It should be noted that all the trading units shown in Figure 7.2 with the exception of the OPEC countries are relatively advanced industrial countries.

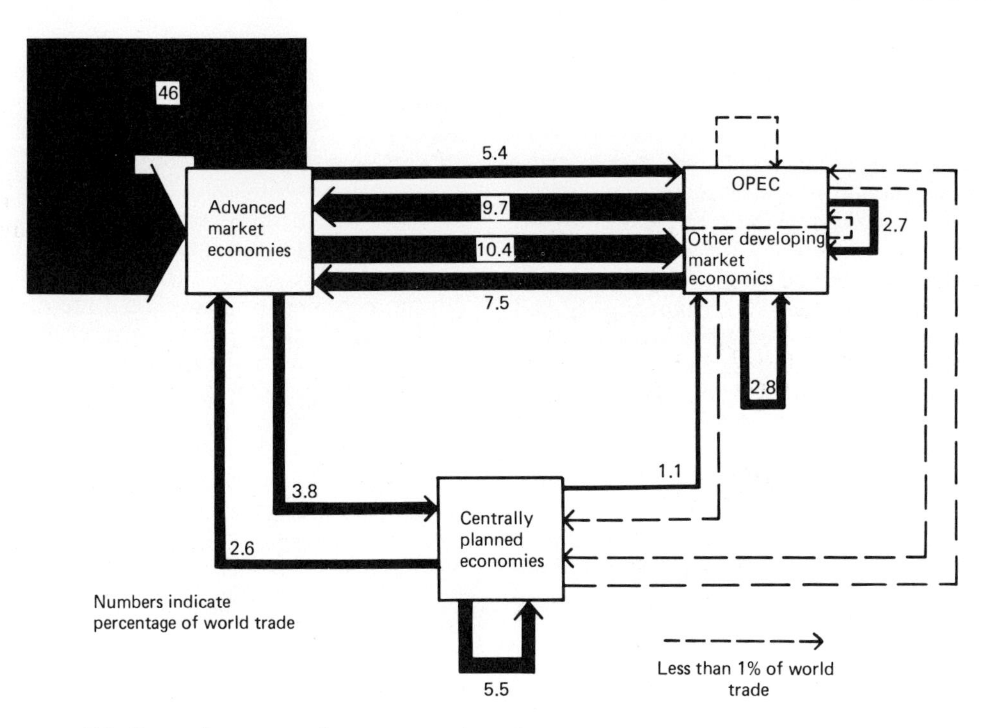

Fig. 7.1 General pattern of international trade

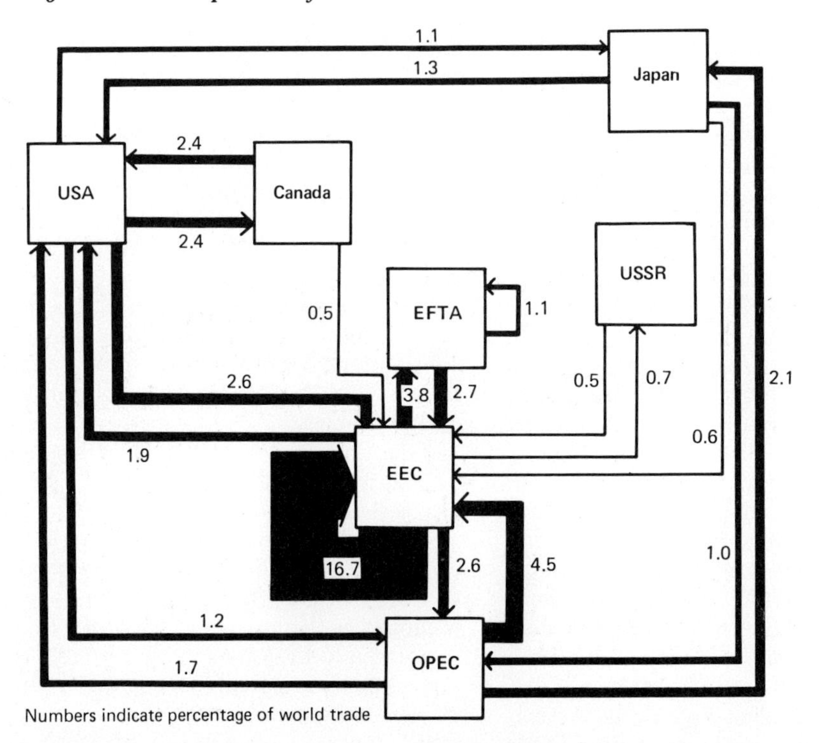

Fig. 7.2 International trade between major trading areas

THE COMPOSITION OF INTERNATIONAL TRADE

Recent trends

In recent years the importance of primary products (food and raw materials) in world trade has declined compared with that of manufactured products, but the importance of the trade in fuels has increased largely through the great development of the trade in petroleum. Manufactured goods have begun to appear prominently even in the exports of developing countries such as Bangladesh, Pakistan and Sierra Leone. Manufactures now dominate the export trade of Hong Kong and South Korea in the Far East.

World trade in food

The general characteristics of international trade in food are illustrated in Figure 7.3(*a*). In this diagram the 45° line shows the positions where the value of food imports is equal to the value of food exports. Any point below this line indicates an area in which food imports are of greater value than food exports. A point above the line indicates that food exports are of greater value than imports.

The advanced market economies are the leading importers of food, but they also export almost as much as they import. This is because they consist of two contrasting groups of countries. On the one hand there are the EEC and Japan which are both net importers of food, then on the other hand there are the advanced countries which have large areas of farm land such as Canada, the USA and Australia and New Zealand which are net exporters of food. The USA, Canada and Australia are responsible for exporting over 80% of the world's wheat which enters trade. Much of this goes to Japan, India and the United Kingdom. The USA also leads in rice exports.

The developing market economies export more food than they import, though the OPEC countries, many of which have desert climates, are net importers of food. India and Sri Lanka are great exporters of tea, mostly to the United Kingdom. Over half of the world's exports of cocoa come from Ghana, Nigeria and Ivory Coast and go mainly to the USA and West Germany. Over one-third of the world's coffee exports come from Brazil and Colombia.

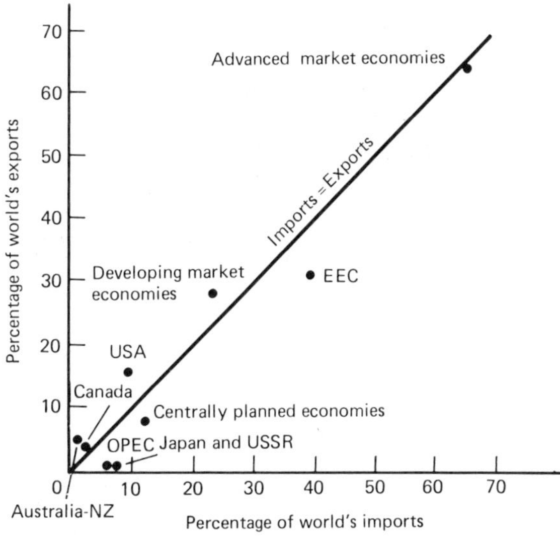

(a) World trade in food

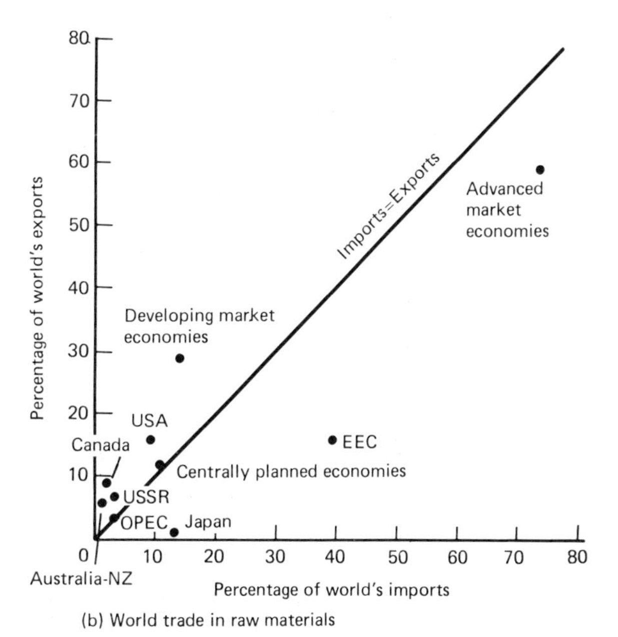

(b) World trade in raw materials

Fig. 7.3 a, b General characteristics of world trade in food and raw materials

World trade in raw materials

Trade in raw materials is illustrated in Figure 7.3(*b*). It follows much the same pattern as the trade in food.

Once again the advanced market economies are both the leading importers and the leading exporters. This time, however, they import considerably more (over 70 %) than they export (under 60 %). This is mainly because the EEC is such a large net importer of raw materials, as is Japan. On the other hand, the USA, Canada, Australia and New Zealand are net exporters. The wool export trade is dominated by Australia and New Zealand, and the declining cotton export trade by the USA. These fibres travel mainly to the EEC and Japan. Australia and Canada also lead in the export of iron ore, mainly to Japan, West Germany and the USA. Japan is unusual in that its trade in raw materials consists almost entirely of imports.

The developing market economies as a whole are net exporters of raw materials. Over half of the world's natural rubber travels from Malaysia, Singapore and Indonesia, mainly to the USA. Chile and Zambia export large amounts of copper.

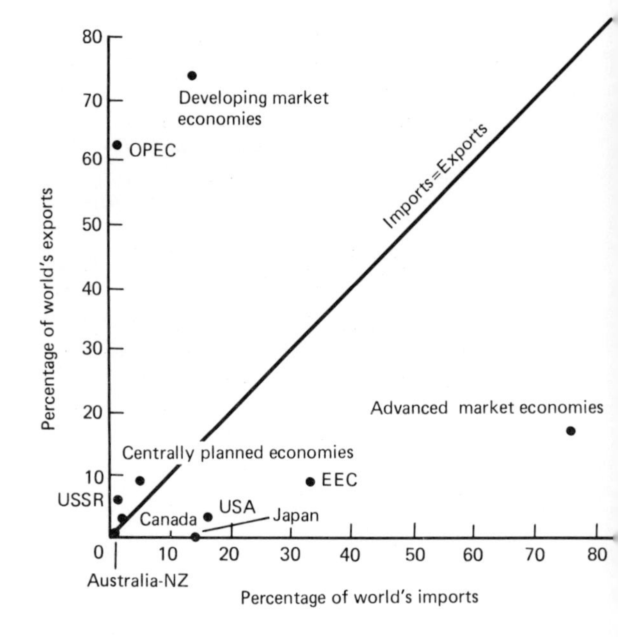

(c) World trade in fuels

World trade in fuels

The fuel trade (Fig. 7.3(*c*)) is now dominated by petroleum. There is a very clear distinction between the developing and the advanced market economies. Over 70 % of the fuel exports come from the developing market economies, mainly in the form of petroleum from the OPEC countries. Well over 70 % of the total exports of fuels go to the advanced market economies, most of this to the EEC, but also large amounts to the USA and Japan.

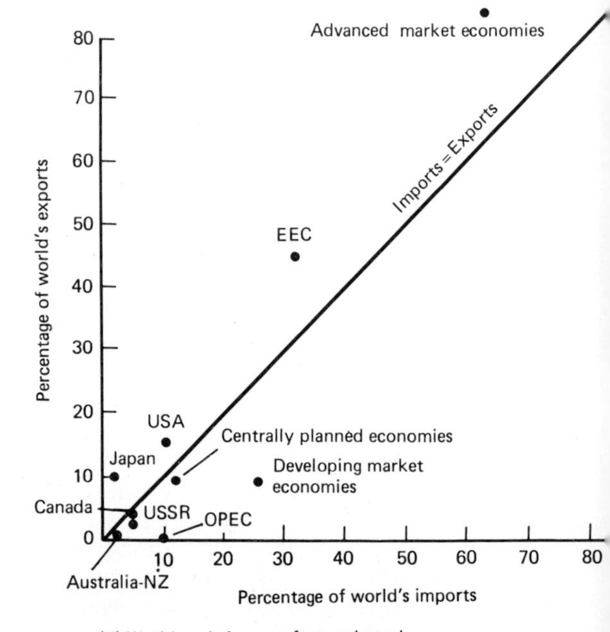

(d) World trade in manufactured goods

Fig. 7.3 c, d General characteristics of world trade in fuels and manufactured goods

World trade in manufactured goods

World trade in manufactured goods (Fig. 7.3(*d*)), is dominated by exports from the advanced market economies such as the EEC, the USA and Japan. The advanced countries also take most of the imports. About one-quarter of the world's imports of manufactured goods are taken by developing countries, but well over 60 % go to the advanced countries. The EEC is both the leading importer and the leading exporter of manufactured goods. Japan is primarily an exporter and takes relatively few imports. Certain small Far Eastern countries have recently become prominent as exporters of manufactured goods. Hong Kong and South Korea are now leading exporters of clothing. West Germany, the USA and Japan lead in the export of motor vehicles. Switzerland handles nearly half of the world's export trade in watches and clocks.

SIMILARITIES AND DIFFERENCES BETWEEN TRADING REGIONS

The general characteristics of the trading regions of the world are shown in Figures 7.4(*a*) and 7.4(*b*). If the world is taken as a whole, 60 % of the imports and exports are manufactured goods, 20 % are food and raw materials and 20 % are fuels. This is shown by the dashed lines on the two triangular graphs. We can read from these graphs the general trading characteristics of the world's main trading regions. It can easily be seen from the graphs that some trading regions in both cases are placed near the world average position, whereas others move towards the corners of the graphs.

The advanced market economies are fairly close to the world average in both graphs. They export rather more manufactured goods and fewer fuels than the world average (Fig. 7.4(*a*)), but their imports are almost exactly the same as the world average (Fig.7.4(*b*)). Most of their trade is with each other rather than with developing market economies or centrally planned economies.

The USA shows a tendency to export a greater amount of both manufactured goods and food and raw materials (Fig. 7.4(*a*)) and to import more than the average proportion of fuels. Its

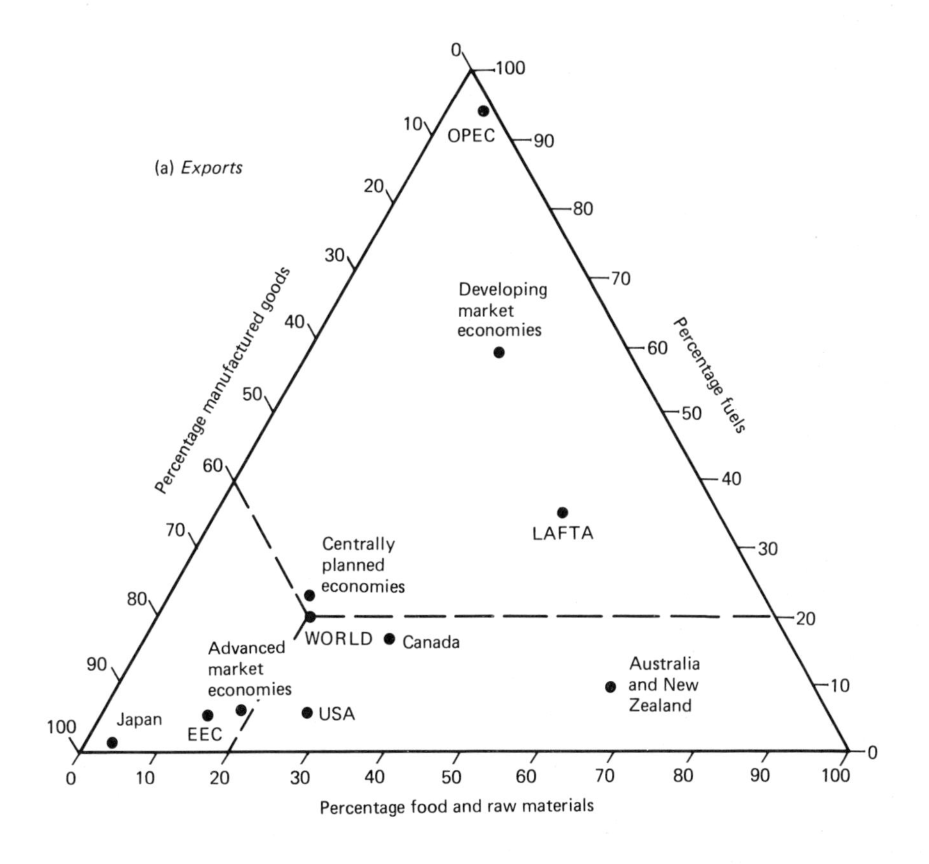

Fig. 7.4 a Exports of world's major trading regions

exports go mainly to other developed countries, particularly Canada and the EEC and a high proportion of its imports come from Canada.

Canada's trading characteristics are quite different. It exports more than the average proportion of food and raw materials but tends to import manufactured goods (70–80% of its imports). Canada's trade is strongly directed towards the USA and, to a lesser extent, towards the EEC.

The EEC concentrates strongly on the export of manufactured goods (80% of the total value of exports). Its main imports are also manufactured goods, but imports of food and raw materials are above the world average. Most of the EEC's trade is between its member countries but it has high levels of exports to EFTA and imports from OPEC.

The centrally planned economies have proportions of exports that are very similar to the world average but they import a larger proportion of manufactured goods and a smaller proportion of fuels. Most of their trade is between their member countries.

In general, all the above trading regions have patterns of exports and imports which are very similar to the world average. On both triangular graphs they are located fairly close to the 'world' position. Some trading regions, however, deviate greatly from the world average.

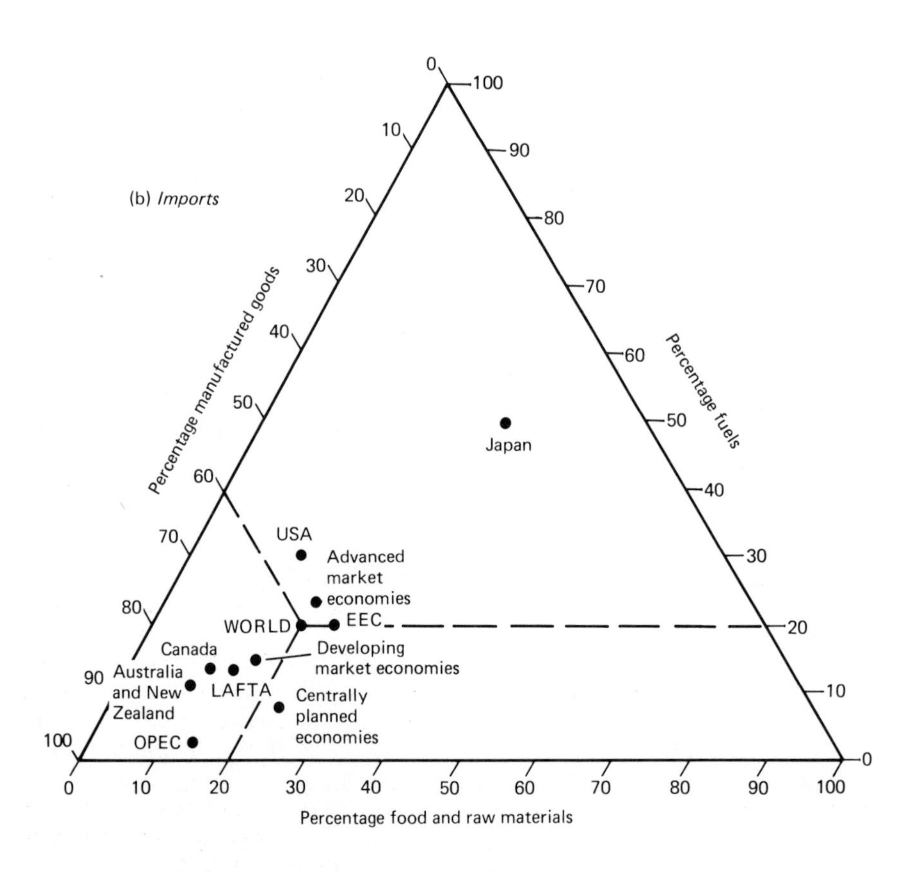

Fig. 7.4 b Imports of world's major trading regions

Developing market economies export very few manufactured goods, and concentrate mainly upon fuels (60 %) and to a lesser extent upon food and raw materials. The OPEC countries are an extreme example, well over 90 % of their exports being fuels. Their import trade is more 'normal', showing only a slight emphasis upon manufactured goods.

Japan also has an unusual pattern of exports and imports. Its exports are almost entirely manufactured goods. About half of its imports are fuels and about a third are food and raw materials. This illustrates the high density of population in Japan and its poverty in natural resources.

Australia and New Zealand contrast greatly with Japan. The majority of their exports are food and raw materials whereas about 80 % of their imports are manufactured goods. The Latin American Free Trade Association (LAFTA) has a trade pattern rather similar to that of Australia and New Zealand. The main differences are that it exports a higher proportion of fuels and imports a rather smaller proportion of manufactured goods. Australia and New Zealand export mainly to Japan and receive most of their imports from the EEC, Japan and the USA. LAFTA trades mostly with the USA and to a lesser extent with the EEC.

7.2 Transport systems of the world

GENERAL EVOLUTION

Since the Middle Ages a number of different stages can be recognized in the development of the world's transport systems.

The pre-railway age

In medieval times transport across land areas was extremely difficult. Roads were generally unpaved and were often quite impassable in bad weather. Most movement over any distance or involving any considerable load was achieved by using horses. The cities of the time were quite small and it was quite possible to move about satisfactorily within them on foot.

The most efficient form of transport was by water, either across the sea or by using rivers and lakes. In North America, for example, the Mississippi river became a major transport route. By 1500 AD long journeys across the ocean had become possible by the use of sailing ships and Europeans began to explore and settle in the other continents. But travel across the oceans involved many dangers.

An improvement began to take place in land transport in the late eighteenth century with the creation of turnpike roads and canals which were able to carry heavy, bulky cargoes.

The introduction of railways and steamships

About the 1830s railways began to be built in Europe and North America. These soon made many canals obsolete since goods and people could be transported so much more quickly. At the same time larger towns developed as they became centres of radiating systems of railway lines.

Also during this period sea transport became much more efficient and reliable and industry could develop in Europe and North America supported by raw materials transported for distances of thousands of miles. Steamships began to replace sailing ships, and ships in general increased greatly in size. This meant that some smaller ports which could not accommodate larger steamships began to decay. In other cases ports adapted themselves to the new conditions by building new, larger docks.

In the early decades of the twentieth century, in Europe and North America particularly, land transport was dominated by the railway, and ocean transport by large cargo ships and passenger liners which could cross the Atlantic in a few days. Also, railways had been introduced to other areas, such as India and Argentina, by European influence.

The modern era

The modern era really began in the 1920s with the growth in importance of transport by motor vehicles and aircraft. The motor car, for the first time, provided individual, as distinct from mass, transport. Also other motor vehicles, such as trucks and buses, began to compete with the railways.

Similarly air transport began to compete with the large passenger carrying ocean liners. Eventually the point was reached when more people came to travel across the North Atlantic by air than by sea. However, air transport cannot compete with

Man's adaptation of nature to create a port: Los Cristianos, Tenerife

sea transport in the bulk transport over great distances of large cargoes of raw materials.

GENERAL CHARACTERISTICS OF THE VARIOUS MODES OF TRANSPORT

Patterns of accessibility

Figure 7.5 illustrates the different kinds of accessibility provided by the different modes of transport.

Road transport is easily the most flexible. Free accessibility is provided to all land areas where roads exist, and vehicles can reach any desired roadside location. Also, road transport can easily adapt itself to changing conditions. If a town expands, for example, roads are naturally created together with buildings, so the new urban areas are automatically accessible to road transport.

Railways, on the other hand, are only accessible to surrounding land areas at passenger stations or goods depots. Between these points, railway trans-

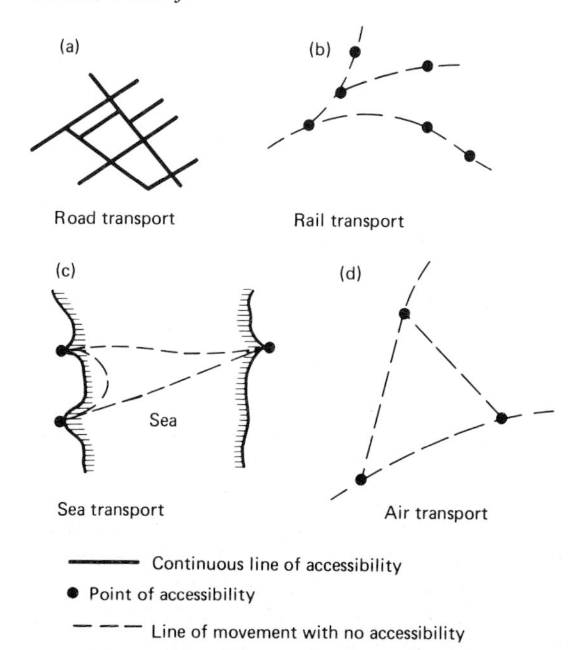

Fig. 7.5 Contrasts in the types of accessibility provided by different modes of transport

port is unavailable, even though one can see the trains going past. Hence, railway transport only provides accessibility at a series of points along lines (Fig. 7.5(*b*)). Also, it is difficult for a railway system to adapt to new conditions. In a growing city, for example, it may not be possible to find space for the building of new railway lines.

Motorways, the most modern form of road transport, are very similar to railways in this respect. In a motorway system accessibility to the surrounding area is available only at the intersections. However, unlike railways, the same vehicle can travel on both the motorway and the ordinary road system.

Canal transport closely resembles railways. Only at certain points can the change from water to road transport be made.

Sea transport (Fig. 7.5(*c*)) involves movement from one specialized port area to another. At these ports, as at railway stations, there has to be a change from one transport mode to another, from sea to either road, rail or canal.

Air transport (Fig. 7.5(*d*)) is very similar to sea transport in this respect. Accessibility is available only at airports. During the journey the aircraft are completely inaccessible, even more so than ships in mid-ocean.

Variations in transport costs

The various transport modes also differ from one another in respect of their costs. Transport costs are unlikely to be exactly proportional to travel distance. It is likely that the average transport cost will decrease as travel distance increases, as shown in Figure 7.6(*b*). This is because of the relationship between fixed (terminal) costs and variable (line-haul) costs (see page 109).

Figure 7.6(*a*) shows how the transport costs of two imaginary modes of transport (modes 1 and 2) vary with distance. It can be seen that mode 2 has a terminal cost of £20 whereas the terminal cost for mode 1 is only £10. These terminal costs remain the same whatever transport distance is involved.

The variable costs for mode 1 increase at a rate of £4 per 100 kilometres whereas mode 2's increase at only £2 per 100 kilometres. Thus, from Figure 7.6(*a*) it can be seen that costs for mode 2 are higher up to a distance of 500 kilometres. At distances greater than this, mode 1's transport costs are higher. Figure 7.6(*b*) shows how the average transport costs of the two modes vary with travel distance. Such average transport cost curves are

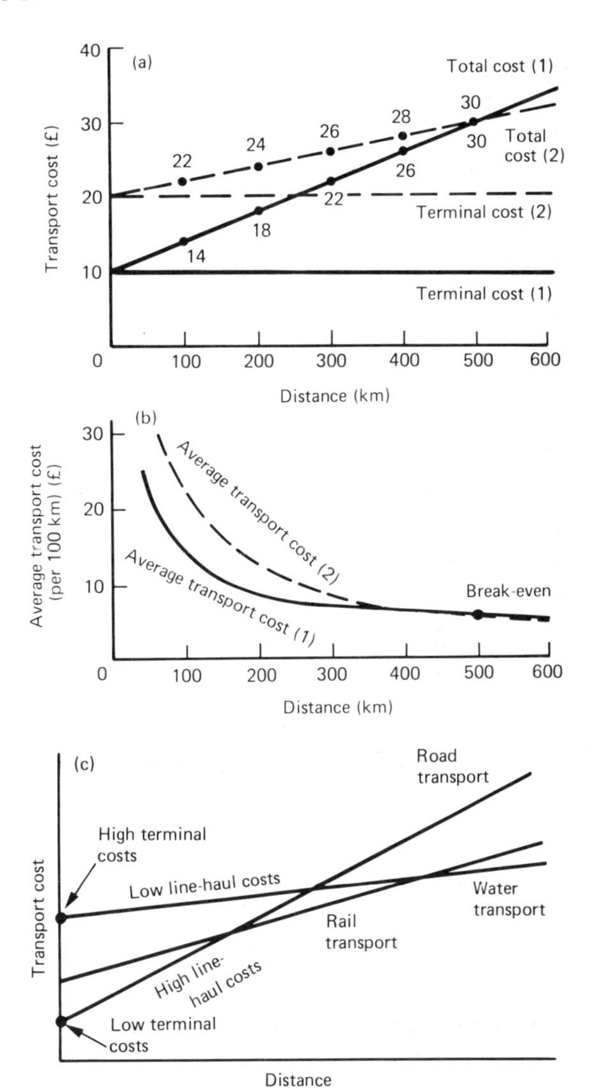

Fig. 7.6 Variations in transport costs

said to 'taper'. They fall quickly at first and then flatten out. The way in which the transport cost curve tapers is governed by the relationship between the terminal costs and the line-haul costs.

Figure 7.6(*c*) shows the general relationship between the transport cost curves for road, rail and water transport. It can be seen that road transport is cheapest for short distances, but rapidly becomes more expensive for longer distances. Water transport is cheapest for longer distances, but is most expensive for short distances. This is because road transport has very low terminal costs (road users do not have to pay for the roads they use) but relatively high line-haul costs. Water transport, on the other hand, has very high terminal costs (the cost of large ships and dock and harbour installations) but very low line-haul costs.

THE CHARACTERISTICS OF THE VARIOUS MODES OF TRANSPORT

Railways

Railways were of outstanding importance during the Industrial Revolution especially in the nineteenth and early twentieth centuries. They were created when cities were much smaller than they are now. Usually a radiating pattern was built around each city and transport to and from the city centre was assisted by bus and tramcar transport to the various stations.

As cities have expanded in size during the twentieth century, and as industries and shops have tended to move from the city centres to the outskirts, the railways have become less useful for movement within large cities. They were usually built to meet the need for radial movements and they are not well suited to movements around the city's circumference.

Some railways were also built for political or economic reasons. The Trans-Siberian railway and east-west transcontinental lines in Canada and Australia were built with the aim of welding their countries together politically. Other railways have been built to link remote mineral deposits to the coast, as in Australia and Canada or to encourage the export of farm products, as in Argentina.

Railways are best suited to linking together two locations by a transport route along which flow bulk shipments of a single commodity such as wheat, coal or iron ore, or even people. For passengers, rail transport begins to suffer competition from air transport when the travel distance rises to about 600 kilometres.

Because railways have existed for so long, and geographical conditions have changed during that time, they have had to try to adapt to new

French Railways' TGV high-speed train

conditions. Unit trains have been developed consisting of a large number of wagons and one or more locomotives, permanently coupled together. These operate rather like a conveyor belt. They run along circular loop lines and may be loaded or unloaded while still moving slowly. Such trains link, for example, electricity generating stations with coal mines, as in Yorkshire (page 94) and iron mines with coastal ports, as in Australia. Very fast passenger trains have been developed to link large conurbations in Japan and also the United Kingdom. These provide direct transport between city centres over distances too short for severe competition from air transport.

The railways have also developed integration with other transport modes. Transhipment from rail to road and sea transport has been eased by the use of containers. Freightliner terminals, to facilitate transfer from road to rail transport, have been established near many cities in Britain.

Water transport

Water transport includes the use of both the sea and inland waterways. Usually there is a need for the transhipment of cargoes when they pass between the sea and inland waterways, because seagoing ships cannot usually travel along canals. But the range of seagoing ships has been extended by the building of ship canals, as between Manchester and the Mersey estuary, and also the more recent creation of the St. Lawrence Seaway by linking the Great Lakes with the St. Lawrence estuary.

Water transport is best suited for transporting bulk cargoes of raw materials and fuel over great distances, when speed is unimportant. Because of its low line-haul costs, water transport is the cheapest way of transporting commodities over very long distances (Fig. 7.6(*c*)).

In recent years there has been a great increase in the size of oil tankers and other bulk cargo vessels such as iron ore carriers. So much so that many ports can no longer accommodate these huge ships and some of them cannot pass through straits such as the English Channel, and the Suez Canal. Special harbour facilities have had to be developed in many countries for these ships.

One of the greatest problems in sea transport

British Petroleum's 250 000 tonne tanker 'British Patience'

has been the slow and expensive loading and unloading of ships at the dockside. During this time, the ship, a valuable capital asset, is lying idle. This problem has been tackled by the development of container and roll-on/roll-off ships. In the former case, cargo is packed in containers of a standard size which can easily be transferred mechanically between land transport and sea transport. In the latter case, road vehicles are carried in ships. Ships now exist which carry loaded barges by sea from one inland waterway system to another, as for example, from the Great Lakes to Rotterdam at the mouth of the Rhine.

The quicker 'turn round' of ships by the use of these methods means that the ship can spend a longer time at sea and therefore it is used more intensively and efficiently. It also means that the need for dock space is reduced. A container port needs a large level area for stacking containers and efficient road and rail links leading inland. Many of the old, cramped dock areas have now become obsolete, and new developments have taken place where more space is available.

Partly for these reasons, the London docks have seriously declined. For many years the trade of the London docks has been moving downstream along the Thames as the increase in the size of ships demanded larger dock areas. Now the abandoned old docks near central London are a serious planning problem. Much of the Port of London's trade, including container handling, roll-on/roll-off facilities and the bulk import of grain and timber, has moved to Tilbury, much nearer the mouth of the Thames. London has also suffered through competition from east coast container ports such as Felixstowe.

Felixstowe was the first container port to be set up in Britain. It lies nearer to the large ports of Belgium and the Netherlands (e.g. Antwerp and Rotterdam) than any other British port. Its efficient dock facilities have attracted trade from all parts of Britain although its immediate hinterland is sparsely populated and has little industry. It is now one of the leading ports in Britain although it was practically derelict in the 1950s.

Road transport

Roads played only a minor part in the Industrial Revolution in the nineteenth century. The road network of that time, used mainly by horse-drawn vehicles, acted as a feeder to the railways and canals. Thus, when motor vehicles began to

The influence of relief on transport routes: the A5 in North Wales

appear on a fairly large scale in the 1920s, they had to move along a transport network which was designed for completely different conditions.

To some extent motor vehicles still have to use this outdated road network, but many improvements have been made which improve the efficiency of the movement of motor vehicles. Major roads have been widened and straightened, and bypasses have been built round small towns and villages.

Motorways were first created in Germany, for strategic reasons, in the period between the two world wars. These provide motor transport with some of the advantages for long distance movement which the railways possessed. The main difference is that road vehicles cannot be assembled into 'trains'. Since the Second World War, Britain and other countries have developed motorway networks which have enormously reduced travel times. At the same time, commercial vehicles have greatly increased in size so as to take fuller advantage of these new facilities for fast, long-distance transport.

The British motorway network is designed to link together the main conurbations such as Merseyside, Greater Manchester and West Yorkshire (M62); Merseyside, Greater Manchester and the West Midlands (M6); West Yorkshire, South Yorkshire, the West Midlands and Greater London (M1); and South Wales and Greater London (M4). In addition, various 'spurs' run to major ports (M62 to Hull), and also to busy holiday areas (M55 to Blackpool, M5 to Devon). In addition a radial system has developed around London. The pattern of motorways is strikingly similar to the pattern of major railway routes.

Road transport is particularly useful for coping with the extremely complex movements which take place in large urban areas, and particularly with the journey to work. Only road transport is flexible enough to meet the needs of the complex interlacing patterns of journeys that have developed as workplaces have located themselves near the outskirts of conurbations as well as in the centre. Also, the road system is available for the general transport needs of business in the period between the morning and evening 'rush hours'.

Air transport

Air transport provides rapid movement between airports in any direction, with few obstacles to movement. This is particularly useful for the transport of people, but it is also an efficient means of transporting goods of small bulk and high value, such as mail and certain perishable goods. As with ocean transport, terminals (in this case airports) have to be provided but, unlike road and rail transport, no transport channel is necessary. Air transport is little affected by physical features of the landscape. Hence it is of particular value in areas like Alaska, where mountains and climate severely hinder surface transport.

In earlier times, when the distance range of aeroplanes was smaller, the world's configuration of land and sea influenced the direction of air routes because of the need for refuelling stops. Thus flights from London to New York might call at Shannon (Ireland) or Gander (Newfoundland). Today, however, non-stop flights are made over great distances, as from Europe to the *west* coast of North America.

This freedom of movement over enormous distances has tended to improve communication between the various national governments of the world. It has become possible for politicians from countries all over the world to meet at conferences in the Caribbean, central Africa, Japan or elsewhere. It has also helped in the management of huge multinational business corporations since their executives have virtually worldwide mobility in a few hours.

In many countries there are great benefits from concentrating international flights at a single airport, with other airports acting as feeders. This is usually the case in Europe where individual countries are fairly small. Heathrow, for example, handles the bulk of international scheduled flights to and from Britain. The USA with its much larger area has a number of extremely large international airports, including Chicago (O'Hare), the busiest of all.

WORLD DISTRIBUTION OF THE VARIOUS MODES OF TRANSPORT

It is possible to classify the countries of the world according to the stage they have reached in the development of their transport system. The developing countries are still in the pre-railway era, as Europe was two hundred years ago. Others have reached the railway age. Railways carry the bulk of their goods and passengers and the ownership of private cars is at a low level. Advanced countries such as those of Europe and

North America, and Japan have developed a much greater variety of transport facilities including air transport, containerization, ocean bulk carriers, etc.

Railway transport

Passenger traffic on railways is generally unimportant in the developing countries of the Far East, Africa and Latin America (pre-railway era), but it has also become relatively less important in Canada and the USA (post-railway era). It is moderately important in most European countries, but its importance rises in the Warsaw Pact countries (Fig. 2.7) and Japan where a high level of efficiency has been achieved.

Since the 1960s railway passenger transport has considerably increased in the developing countries. In parts of Africa, such as Morocco, Algeria and parts of West Africa, passenger transport has doubled in volume in 10 years.

In Canada and the USA there has been a considerable decline in passenger traffic but in Europe there has been a general but variable increase. The greatest recent increases have been in the less developed countries such as Greece, Portugal, Bulgaria and Romania. More advanced countries such as Belgium, Denmark, West Germany and Switzerland have had a lower rate of increase.

Goods traffic is not very important in most of the developing countries but it may rapidly gain in importance with the discovery and development of a new mineral resource. Goods traffic is of considerable importance in most advanced countries, including the USA and Canada, where long-distance transport of bulk commodities is needed. In Canada goods traffic is increasing in importance. In Europe goods traffic is generally less important, but its importance increases in southern and eastern Europe where, in some countries, it is expanding rapidly. There has been a great increase in goods traffic recently in Australia, with the opening up of mineral deposits, but in Japan it is declining.

Water transport

Measured by the weight of goods transported, sea-borne trade is highly concentrated in particular countries. Of greatest importance are the oil and mineral producers such as Venezuela and the Middle Eastern countries, and Australia. Also advanced oil and mineral importing countries such as the USA and some western European countries have high volumes of sea-borne trade.

In Europe, the Netherlands handles an enormous amount of sea-borne trade in relation to its size and population. The weight of goods entering and leaving the Netherlands by sea transport is more than double that of West Germany. This reflects the importance of Rotterdam as a port which serves parts of West Germany as well as the Netherlands.

Road transport

The USA possesses about 40% of the world's motor cars. This amounts to about one car for every two people in the USA. Canada and Sweden have at least one car to every three people, and the more advanced countries of Europe (the United Kingdom, France, Belgium, the Netherlands, West Germany, Denmark, Norway, as well as Italy and Iceland) have more than one car to every four people. In recent years the rate of increase in car ownership has tended to be greatest in Asia, and also in the European countries with a relatively low level of car ownership, such as Czechoslovakia, Greece, Hungary, Poland, Spain and Yugoslavia.

The USA possesses almost 40% of the world's commercial vehicles and the advanced countries of Europe also have large numbers. The less developed parts of the world have fewer commercial vehicles than the advanced countries, but they usually have a higher proportion of commercial vehicles than of motor cars. Recent rates of increase in the number of commercial vehicles have been greatest in the less developed countries, and particularly in Africa. In Europe these rates of increase have been greatest in the less developed countries of eastern and southern Europe. It appears that, as countries develop economically, an expansion in numbers of commercial vehicles takes place at an earlier stage than the expansion in passenger motor cars.

Air transport

Air transport of passengers is strongly concentrated in North America, and most of this consists of movement *within* either the USA and Canada. These countries are large enough for air transport to be a feasible alternative to land transport, particularly for passengers who are anxious to reduce their travel time.

Passenger traffic also reaches a high level in

Europe, but here movement is mainly international. Individual countries are generally too small to justify very many domestic flights. In contrast, in the USSR, where passenger air traffic is almost at the level of Europe, almost all movement is internal. In Europe the air transport of passengers is particularly important in the Scandinavian countries such as Norway and Sweden, which have large northern areas with difficult relief and climatic conditions. Iceland, too, is highly dependent upon air transport.

The transport of goods by air has increased considerably in recent years. It is strongly concentrated in North America, Europe and the USSR. To a greater extent than passenger transport it tends to flow from country to country, except in the USSR where it remains mostly internal. Compared with passenger traffic, the air transport of goods tends to be particularly important in Europe (particularly in West Germany, Belgium and the Netherlands) and South America.

7.3 Transport routes and networks

THE LOCATION OF A TRANSPORT ROUTE

Suppose that a transport route is to be created to link together two towns. Certain costs will be incurred. The land will have to be bought and then there will be the cost of actually building the road or railway and of operating the vehicles which use the route. These costs will tend to increase as the length of the route increases. Once created, the transport route may begin to earn revenue as vehicles begin to use it. Some transport routes may be designed to keep costs to a minimum. Others may be designed to make the maximum profit.

The minimization of costs

Crossing a barrier is illustrated in Figure 7.7(*a*). The transport route is to be created to link points A and B. Normally one would expect the cheapest route to be route 1, the straight line linking the two points. In this case however a barrier exists between A and B. This barrier may be a mountain range or an area of swampland, or the estuary of a river. In respect of sea transport the barrier may be a long peninsula, and the building of a canal across it is being considered.

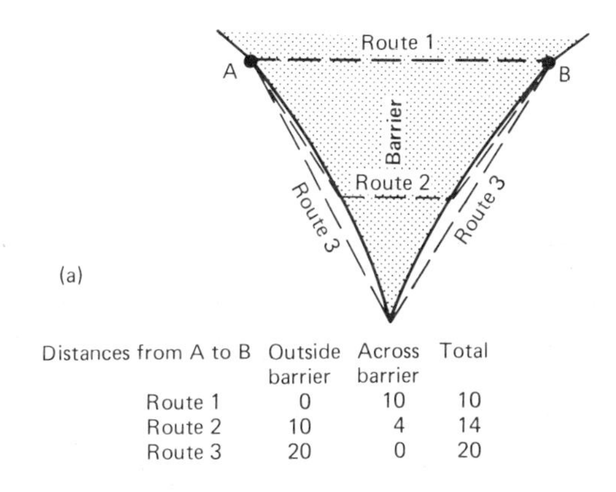

Distances from A to B	Outside barrier	Across barrier	Total
Route 1	0	10	10
Route 2	10	4	14
Route 3	20	0	20

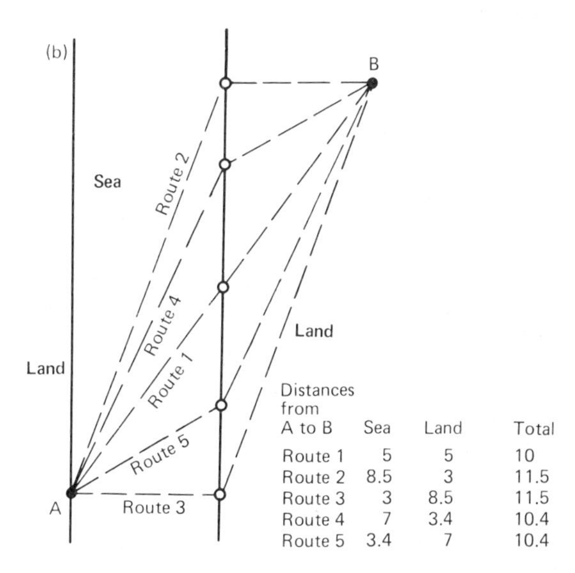

Distances from A to B	Sea	Land	Total
Route 1	5	5	10
Route 2	8.5	3	11.5
Route 3	3	8.5	11.5
Route 4	7	3.4	10.4
Route 5	3.4	7	10.4

Fig. 7.7 Least cost routes across a barrier

The problem is to find the least-cost route. This all depends on the expense of building the route across the barrier. If there is no extra cost involved in crossing the barrier then the cheapest route will be route 1 because it is the shortest.

Suppose that the cost of crossing the barrier is £1.50 per unit distance compared with £1 outside the barrier. Route 1 is still the cheapest route (10 × £1.50 = £15). Route 2 costs £16 (10 × £1 + 4 × £1.50) and route 3 costs £20 (20 × £1).

If the cost of crossing the barrier rises to £2 per

unit distance, route 2 becomes the cheapest because of its shorter crossing of the barrier (10 × £1 + 4 × £2 = £18). Route 1's cost has risen to £20 (10 × £2) and route 3, as before, costs £20.

If the cost of crossing the barrier rises to £3 per unit distance, route 3 becomes the cheapest (£20, as before). Route 1 now costs £30 (10 × £3) and route 2 costs £22 (10 × £1 + 4 × £3).

Illustrations of this principle are the Panama and Suez Canals. Before they were built the cheapest way to travel from the east to the west coast of North America was via Cape Horn, and the cheapest route from England to India was via the Cape of Good Hope. No such 'short cut' canal has been built across the Malaysian peninsula between India and China.

Crossing the sea is illustrated in Figure 7.7(*b*). The problem is to discover which of the five routes from the port A to the inland city B is the cheapest with various transport costs across the sea and the land.

If transport costs across both the sea and the land are the same, then route 1 is the cheapest because it is the shortest route. If all transport costs are £1 per unit distance, route 1 costs £10 which is lower than all the others. If transport across the sea is completely free, the cheapest route is route 2 because, along this route, the shortest possible distance is travelled across the land. Similarly, if land transport is free, route 3 is the cheapest.

Suppose now that sea transport is relatively cheaper than land transport. Sea transport costs £1 per unit distance whereas land transport costs £2. Route 4 now becomes the cheapest (7 × £1 + 3.4 × £2 = £13.8), followed by route 2 (8.5 × £1 + 3 × £2 = £14.5). These two routes travel further across the sea (at the cheaper rate) than across the land. They are both cheaper than route 1 even though they are longer. This 'bending' of transport routes is called 'refraction'.

In the same way, if land transport were cheaper than sea transport we should find that routes 3 and 5, which pass long distances over land, would be the cheapest.

An illustration of this principle is the case of oil from the Middle East to Europe being sent in huge tankers (at very low transport cost) all the way round the southern tip of Africa instead of by the much shorter overland route. These very large tankers are not able to pass through the Suez Canal.

The maximization of profits

The maximization of profits is illustrated in Figure 7.8. A transport route is to be built from town A to town H which will cost 12 units if it is built in a straight line from one town to the other. Since the revenue from these towns will be 20 (10 + 10), the profit from this route will be 8 (20 − 12). In the same region there are other towns, B, C, D, E, F and G. Their locations and the amount of revenue they will produce are shown in Figure 7.8.

If the transport route is designed so as to link together all these towns (Fig. 7.8(*a*)) the profit received is 10; but it is clearly possible to improve on this profit level. In Figure 7.8(*a*) it is clearly

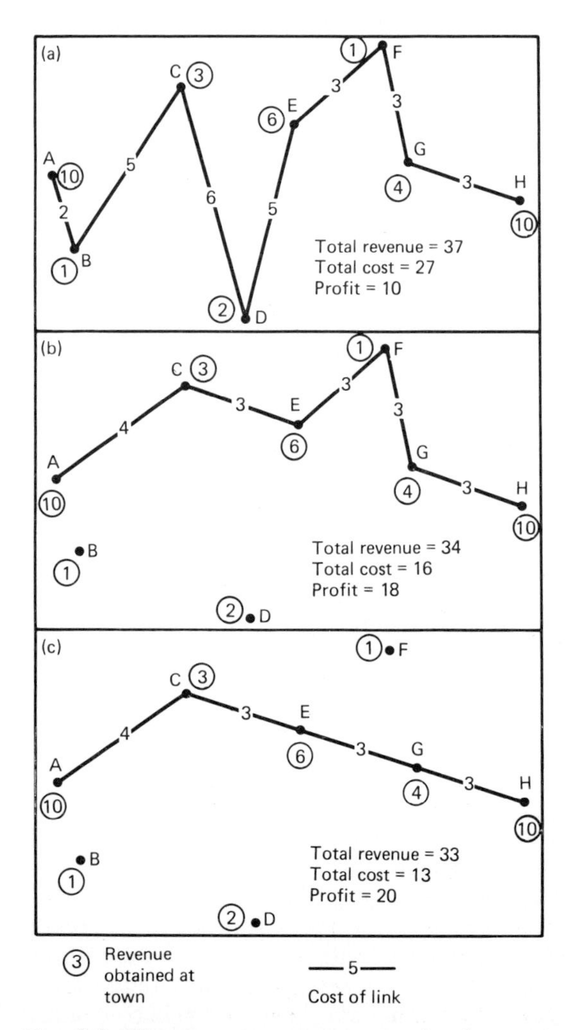

Fig. 7.8 Planning a transport route for the maximization of profits

unprofitable for the route to go to town B. A direct route from A to C costs 4 (Fig. 7.8(*b*)), whereas a route via B costs 7 (Fig. 7.8(*a*)). The revenue from town B is only 1. Therefore profit is reduced by 2 as a result of passing through B. Similarly, profit is reduced by 6 as a result of passing through D. Hence the new route, ACEFGH (Fig. 7.8(*b*)) increases profit by 8.

It is now clear that it is also unprofitable to link with town F. This route costs an extra 3 compared with a direct route from E to G, and it only produces 1 unit of revenue. Hence 2 more units of profit are gained by omitting town F.

In Figure 7.8(*c*) it would not increase profit to omit town C. The saving in costs would be less than 1 unit, and 3 units of revenue would be sacrificed. Hence, ACEGH is the profit-maximizing route.

THE CHARACTERISTICS OF TRANSPORT NETWORKS

A transport network consists of a number of linked transport routes. As well as the transport routes, a network also contains nodes or vertices which are usually the settlements served by the routes. A junction of routes is also regarded as a node or vertex. Also a node (vertex) exists at the end of any route which forms a cul-de-sac.

Builder and user costs

In a least cost to builder network the total length of the routes that make up the network is as small as possible.

In a simple two node case (Fig. 7.9(*a*)) it is easy to see that a straight route joining the two nodes is the shortest way of linking them together.

When there are three nodes the problem is not quite so simple. If the three nodes are arranged in a straight line, or so that the lines linking them form an angle of at least 120° at the central node, the least cost to builder network consists of two routes running directly from node to node (Fig. 7.9(*b*)). If the three nodes are situated at the vertices of a triangle of which all the interior angles are less than 120°, the routes do not run directly from node to node but meet at the point inside the triangle where they form three equal angles of 120° (Fig. 7.9(*c*)). Figure 7.9(*d*) shows a case in which there are five nodes. It is still dominated by 120° angles.

If the builder reduces his costs, and makes the network as short as possible, he increases the cost

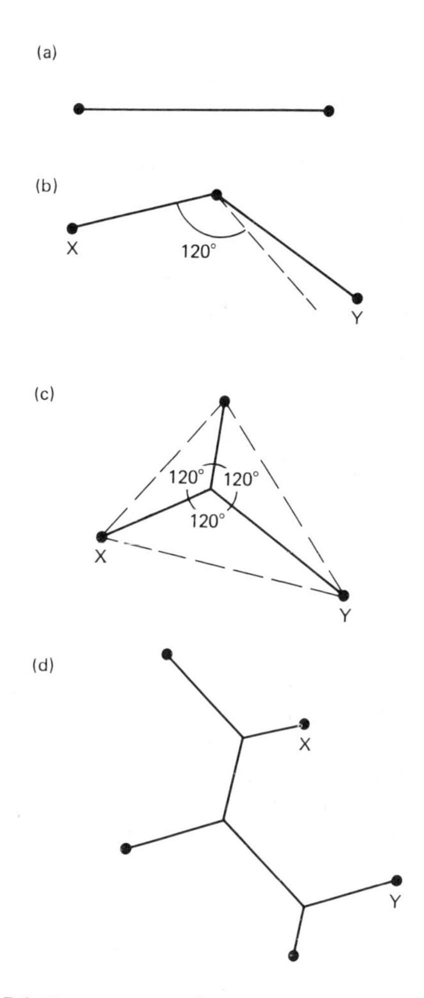

Fig. 7.9 Least cost to builder networks

to the user by compelling him to travel further than is necessary. The 'detour index' measures the directness of travel between a pair of nodes in the network. It is calculated by dividing the network distance between two nodes by the straight line (desire line) distance. The detour indices for the journeys from X to Y in each of the networks in Figure 7.9 are: 1.00 (in (*a*)), 1.10 (in (*b*)), 1.16 (in (*c*)) and 1.98 (in (*d*)). In network (*d*), the network distance from X to Y is almost double the desire line distance.

In a least cost to user network the user of the network is able to travel from one node to another as directly as possible. Ideally his network distance is the same as his desire line distance, and the detour index is 1.00. Some examples of least cost to user networks are given in Figure 7.10.

Least cost to builder networks are most likely to be found in sparsely populated areas where traffic densities are relatively light. The railway networks of northern Canada and the interior of Australia are of this type. They are also found in areas of difficult relief. In the Scottish Highlands, roads are channelled along the valley floors and it is almost impossible to establish links across the mountains. Most of the English motorway system is an example of a least cost to builder network, because motorways are so expensive to build that it is impossible to duplicate routes. If one travels solely by motorway therefore one has to tolerate long detours. A motorway journey from Manchester to Sheffield, for example, is via Leeds and is about 50% longer than the route through the Peak District.

Least cost to user networks are found most commonly in the great conurbations, where the high volume of traffic justifies the creation of a very dense road network. In such areas there are many alternative routes for any journey. These networks are also found in fairly densely populated rural areas, like parts of East Anglia, where a great many roads and lanes have been created through the centuries to link up the villages.

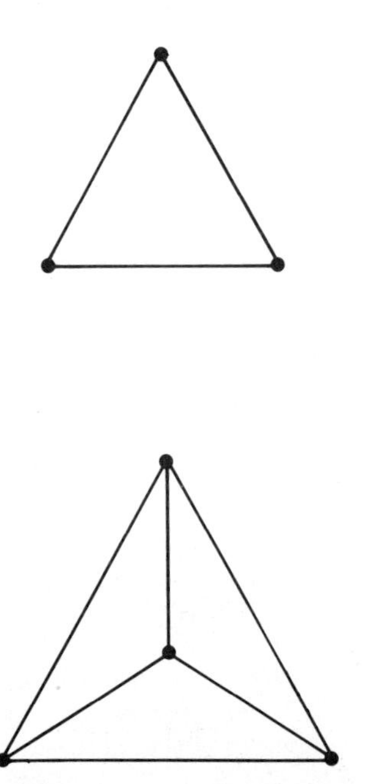

Fig. 7.10 Least cost to user networks

Topological characteristics of networks

In studying transport networks we use a type of geometry known as topology, which is concerned with the relative positions and the relationships between points and lines. The transport network is changed into an abstract form known as a graph. We shall only consider *planar graphs*, i.e. graphs in which all the vertices and edges are in the same plane. If two edges cross they must form a vertex where they cross. One cannot pass over or under the other.

The main topological characteristics of networks are:

(*a*) Elements of graphs (Fig. 7.11)

(i) *Vertices* may be towns or villages or the junctions of edges.

(ii) *Edges* are the lines (routes) that link the vertices together. Only one edge may link together any pair of vertices. An edge forming a cul-de-sac is called a *branch*.

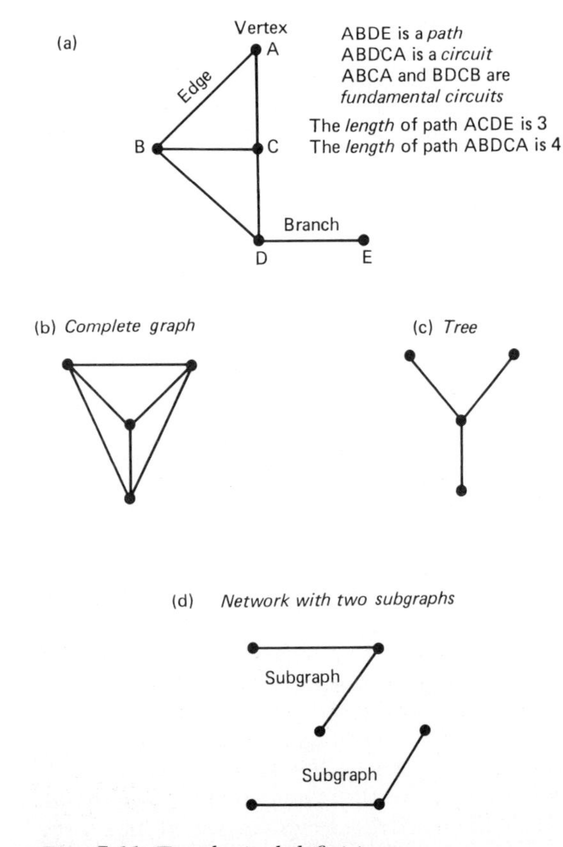

Fig. 7.11 Topological definitions

(iii) A *path* is a sequence of edges linking several vertices in turn. The *length* of a path is the number of edges in it.
(iv) The *topological distance* between two vertices is the length of the shortest path between them.
(v) A *circuit* (closed path) is a path that begins and ends at the same vertex but never crosses its own route nor passes along the same edge twice.
(vi) A *fundamental circuit* is a circuit that does not enclose any other circuits.

(*b*) Types of graphs (Fig. 7.11)
(i) A *connected graph* is one in which every vertex is linked to the network.
(ii) A *complete graph* is one in which every vertex is linked directly to every other vertex.
(iii) A *tree* is a connected graph with no circuits.
(iv) A *subgraph* is a detached part of a network. A network may consist of two or more separate subgraphs.
(v) A *directed graph* is one in which movement along an edge or edges is only possible in one direction. Usually we assume that two-way movement is possible in topological graphs.

The connectivity of networks
The connectivity of a network is the degree of completeness of the linkages (edges) between the various vertices.

The minimum number of edges needed to link together all the vertices of a network is one less than the number of vertices. This results in the formation of a tree (with no circuits). If any extra edges are added, circuits must be formed. This rule can be written as $e_{min} = v - 1$.

The maximum number of edges that can exist in a network without duplication of linkages is found by subtracting two from the number of vertices and then multiplying by three. This rule can be written as $e_{max} = 3(v-2)$.

The *beta index* is calculated by dividing the number of edges by the number of vertices. It can be written as e/v. The higher the value of the beta index the greater the connectivity of the network. Trees and other simple networks have a value of less than 1 (meaning that there are fewer edges than vertices). In a tree the number of edges is $v-1$ (see above), i.e. one less than the number of vertices. Hence the beta index of a tree must be less than one (Fig. 7.12(*a*)). A connected network with one fundamental circuit has a beta value of 1. In this case $e = v$; hence there is an extra edge in addition to those needed to form a tree (Fig. 7.12(*b*)). If there are more than one fundamental circuits then e will be greater than v. Hence the beta value will be greater than one (Fig. 7.12(*c*)).

The more edges a graph has, the more fundamental circuits it has (Fig. 7.12). If we subtract $v-1$ (the number of edges needed to form a tree) from e (the number of edges a network has), this must give us the number of fundamental circuits the network has. Thus, for a connected graph, the

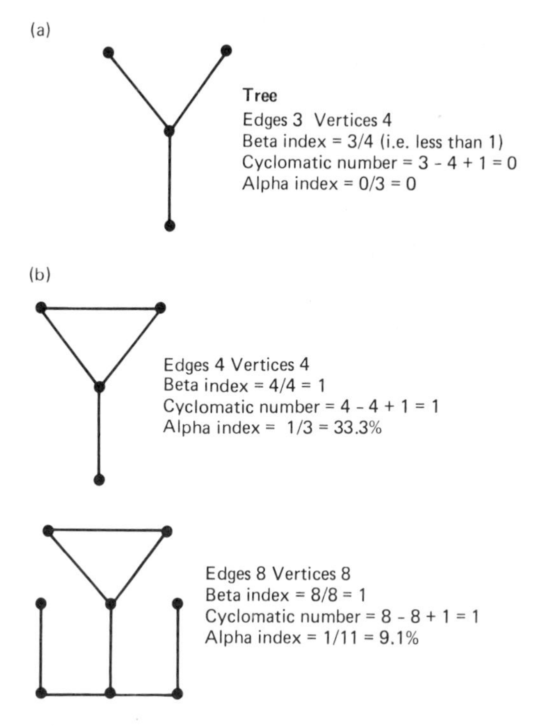

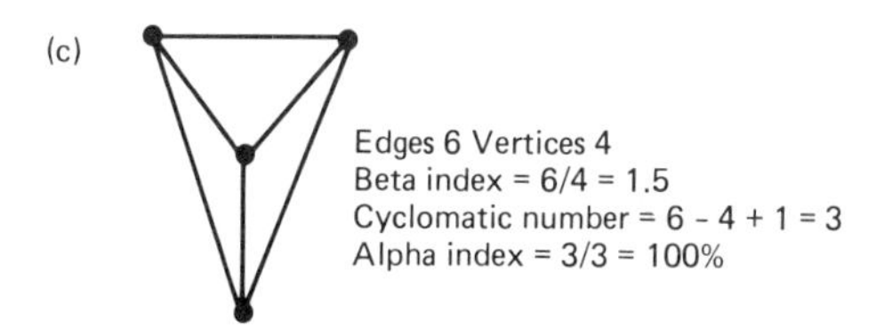

Fig. 7.12 Connectivity indices

number of fundamental circuits is $e-(v-1)$ which is equal to $e-v+1$, which is the *cyclomatic number*. If the network has subgraphs, the number of subgraphs is substituted for the 1 in this expression. In practice, it is easiest to find the cyclomatic number simply by counting the number of fundamental circuits in the network. Unfortunately, the cyclomatic number is not a very good indicator of connectivity unless it is applied to networks that all have the same number of vertices. Two networks may have the same cyclomatic number and yet have very different levels of connectivity. The larger network's connectivity will be smaller. This is explained in Figure 7.12(*b*).

The *alpha index* overcomes this problem of the cyclomatic number by expressing the number of fundamental circuits as a percentage of the maximum possible number for the particular network. It is therefore calculated by dividing the number of fundamental circuits $(e-v+1)$ by $2v-5$ (the maximum possible number of fundamental circuits for the network). The result may be expressed as a percentage. The values of the alpha index range from 0 to 1 (or 0 to 100%). A value of 0 indicates that the network has no fundamental circuits (Fig. 7.12(*a*)). A value of 1 in a connected graph indicates that the network possesses the maximum possible number of fundamental circuits and edges, and that it has the maximum possible degree of connectivity (Fig. 7.12(*c*)).

A similar, but somewhat inferior, measure of connectivity to the alpha index is the *gamma index* which relates the number of edges a network possesses to its maximum possible number of edges (e divided by $3(v-2)$).

Accessibility in networks

The pattern of linkages between the vertices of a network provides a certain degree of accessibility in respect of each individual vertex and also in respect of the network as a whole. A high level of accessibility means that it is possible to travel from vertex to vertex by the shortest possible route.

A shortest path matrix shows the topological distance (measured in edges) (page 143) between each pair of vertices along the shortest path between them. Figure 7.13(*a*) shows a network and its related shortest path matrix. In the matrix, as well as in the network it can be seen that four edges separate A from B.

When we assume, as in this case, that two-way movement is possible along each edge, the shortest path matrix is symmetrical. The numbers to the upper right of the blank diagonal are the exact mirror image of those to the bottom left. If the network were of the 'directed' type (page 143), and movement were only possible in one direction along some of the edges, the shortest path matrix would be asymmetrical.

The accessibility of each vertex is indicated by the Shimbel index and the associated number. The *Shimbel index* is the sum of the distances from any single vertex to all the other vertices. In the shortest path matrix, there is a Shimbel index for the row belonging to each vertex. The *associated number* is the distance from each vertex to the vertex which is most distant from it. The sum of all vertices' Shimbel indices gives the *dispersion index*, which is a measure of the level of accessibility provided by the network as a whole.

The basic factor influencing the level of accessibility in a network is the connectivity of that network. If the network has a high level of connectivity, a large number of edges in relation to the number of vertices, it is obvious that paths from vertex to vertex will be generally direct and relatively short. An example is shown in Figure 7.12(*c*). Each vertex is only one edge distant from all the others.

The second factor is the shape of the network. Figures 7.13(*a*) and 7.13(*b*) show two networks (trees) of exactly the same level of connectivity, but the edges are arranged in different patterns. The shortest path matrices show that network (*b*) provides a higher level of accessibility than network (*a*). In Figure 7.13(*c*), an extra edge (BD) has been added to network (*b*). This has reduced the dispersion index from 58 to 52, but this improvement in accessibility has not been shared equally between the vertices. Vertex D has derived the greatest benefit, its distance to A, B and C having been reduced. Vertices A, B and C have benefited only in respect of their distance from D, and vertices E and F have not benefited at all.

Interaction and traffic flows in networks

The function of networks is to carry traffic of one kind or another. This may be road or rail traffic, telephone calls, letters, or even electric current. In this case we are supposing that the vertices are towns and the edges are roads. Traffic originates at each of the towns and travels along the roads to

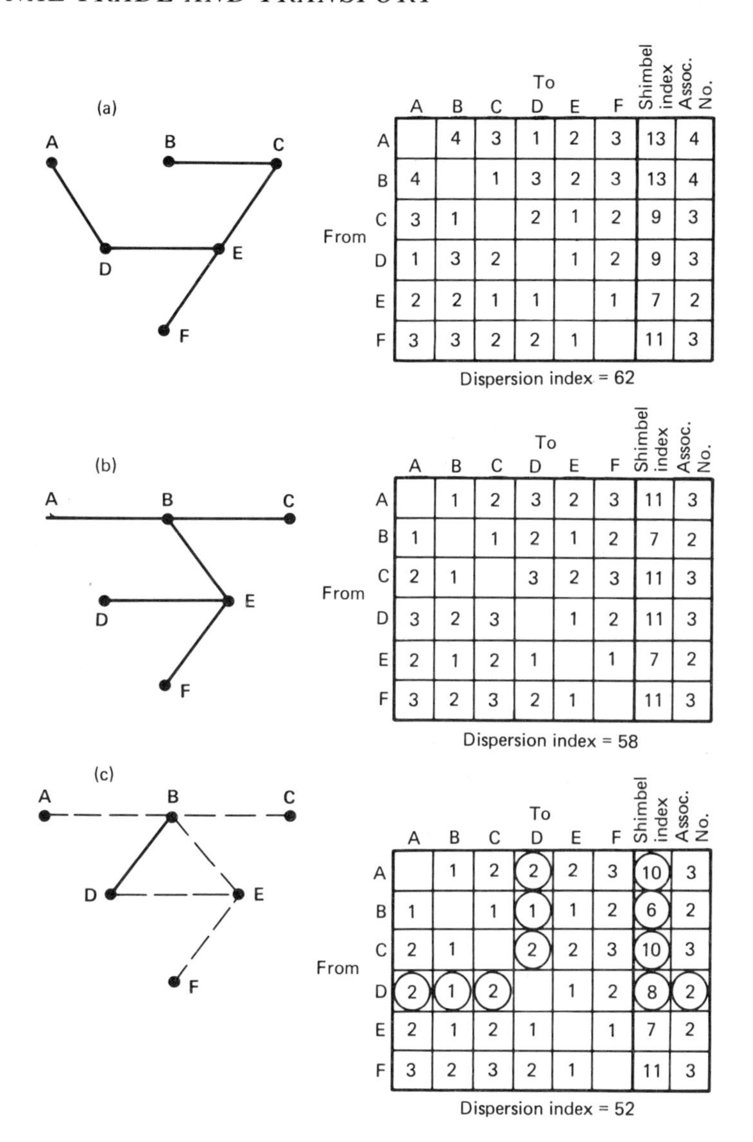

(a)

From \ To	A	B	C	D	E	F	Shimbel index	Assoc. No.
A		4	3	1	2	3	13	4
B	4		1	3	2	3	13	4
C	3	1		2	1	2	9	3
D	1	3	2		1	2	9	3
E	2	2	1	1		1	7	2
F	3	3	2	2	1		11	3

Dispersion index = 62

(b)

From \ To	A	B	C	D	E	F	Shimbel index	Assoc. No.
A		1	2	3	2	3	11	3
B	1		1	2	1	2	7	2
C	2	1		3	2	3	11	3
D	3	2	3		1	2	11	3
E	2	1	2	1		1	7	2
F	3	2	3	2	1		11	3

Dispersion index = 58

(c)

From \ To	A	B	C	D	E	F	Shimbel index	Assoc. No.
A		1	2	2	2	3	10	3
B	1		1	1	1	2	6	2
C	2	1		2	2	3	10	3
D	2	1	2		1	2	8	2
E	2	1	2	1		1	7	2
F	3	2	3	2	1		11	3

Dispersion index = 52

Fig. 7.13 Networks and shortest path matrices

the other towns.

The *gravity model* can be used to estimate the interaction between pairs of towns. This interaction will be represented by flows of traffic. Thus, the gravity model can be used to estimate probable flows of traffic between pairs of towns.

The gravity model, used in this way, takes account of two principles. The first of these is that larger towns will generate more traffic than smaller towns. The second is that more traffic will travel between towns that are situated close together than between towns that are far apart. These two principles are built into the standard equation which is used to estimate the flow of traffic between a pair of towns:

$$\text{Interaction (between Towns A and B)} = \frac{\text{population of Town A} \times \text{population of Town B}}{(\text{distance between Towns A and B})^2}$$

In the denominator the distance exponent is here given as 2. Some studies have used different distance exponents, depending on the type of transport being considered. With air transport, for example, distance will be less of a barrier to interaction, so a smaller exponent is used. If only foot transport is available, distance will severely limit movement, so a much greater exponent is used.

The gravity model will not automatically produce an estimate of the *actual* number of vehicles that move from town to town, but it will allow comparisons to be made of estimated traffic flows between various pairs of towns. When the gravity model is applied to a transport network the various calculations of comparative traffic flows between pairs of towns are set out in an interaction matrix (Fig. 7.14(*b*)).

Figure 7.14(*a*) shows the transport network which is also shown in Figure 7.13(*c*). In this case however, the population total of each town is given. Figure 7.14(*b*) shows the interaction index for each pair of towns and the total interaction between each town and all the other towns.

In Figure 7.14(*c*) all the values given in the interaction matrix have been allocated to the various links of the network, assuming that movement from town to town is always by the shortest possible route. Links AB, BC and EF are alike in that they each lead to a single town. Hence each of these carries the total interaction between that town and all the other towns in the network. This is the figure for 'total interaction' in Figure 7.14(*b*). The other links are more complex. BD carries traffic only between town D and each of towns A, B and C. Link BE carries the traffic of towns. A, B and C to and from towns E and F. Link DE only carries traffic between town D and towns E and F. The greatest traffic flow is along link BE which joins the two largest towns.

Figure 7.14(*d*) is a 'gravity potential' map. The number against each town is its total interaction

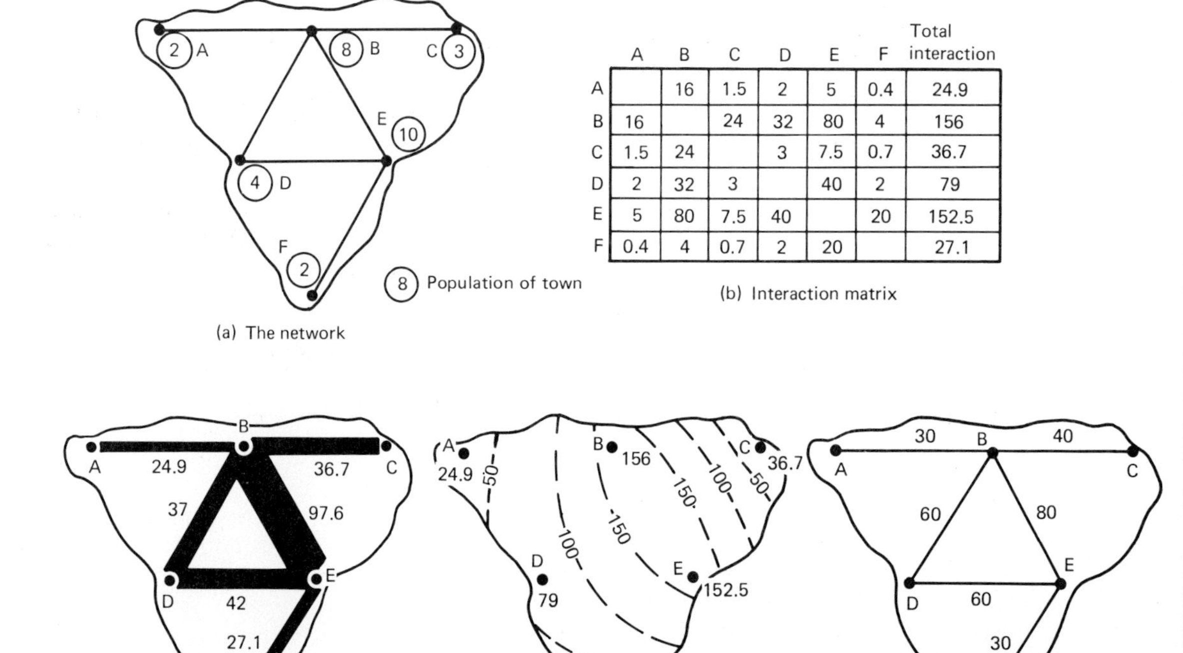

	A	B	C	D	E	F	Total interaction
A		16	1.5	2	5	0.4	24.9
B	16		24	32	80	4	156
C	1.5	24		3	7.5	0.7	36.7
D	2	32	3		40	2	79
E	5	80	7.5	40		20	152.5
F	0.4	4	0.7	2	20		27.1

Fig. 7.14 The application of the gravity model to a transport network

with all the other towns. Isopleths have then been drawn to show the variations in total interaction over the area of this island.

Suppose the carrying capacity of the various transport links is as shown in Figure 7.14(*e*). All the links are able to carry the traffic flows shown in Figure 7.14(*c*), except for link BE which has a capacity of only 80 and is required to carry a flow of 97.6. The solution is for the excess traffic (17.6) to travel via D. This raises DE's flow to 59.6, which it can just handle, and BD's flow to 54. 6 which it can also handle.

It is fairly easy to find the maximum possible flow of traffic between two points on a transport network. In Figure 7.15, for example, the maximum possible traffic flow from A to B is equal to the sum of the capacities in the 'minimum cut'. A 'cut' is any collection of links which completely separates A from B. Four 'cuts' are shown in Figure 7.15, but there are several others. In this case the maximum traffic flow from A to B is 9, because only 9 can cross cut 4.

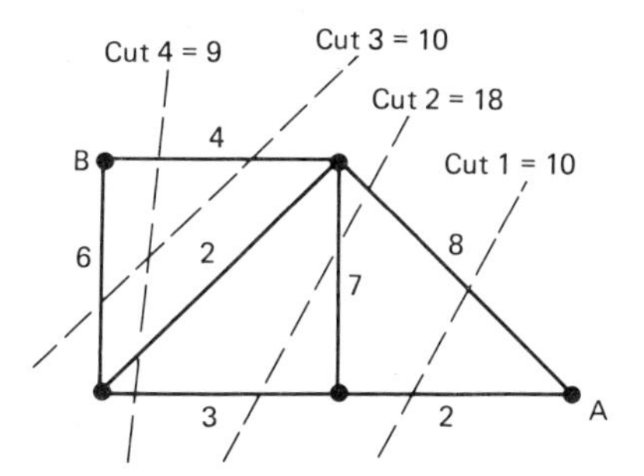

Fig. 7.15 Maximum traffic flow through a network

Exercises

1. (*a*) Discuss the statement that world trade takes place predominantly between the advanced countries of the world rather than between the advanced countries and the developing countries.
(*b*) In which commodities are the developing countries particularly involved in world trade?

2. Discuss the factors that appear to influence the comparative importance of road, rail, water and air transport in different countries.

3. Describe and evaluate the various factors that may influence the commercial viability of
(*a*) an airport;
(*b*) a seaport;
(*c*) a high-speed railway route;
(*d*) an inland waterway.

4. Explain the reasons why so small a proportion of world trade exists between the developing countries.

5. Discuss the various factors that may influence the comparative volumes of traffic flows along the various links of a transport network.

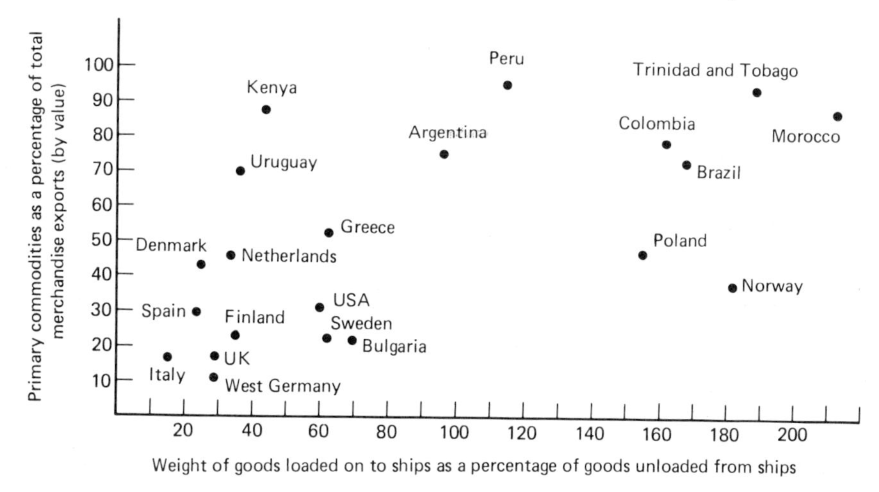

Fig. 7.16 Some characteristics of foreign trade

6. Explain the relationships in a transport network between

(*a*) minimization of costs and maximization of profits;
(*b*) mode of transport and transport costs;
(*c*) builder costs and user costs;
(*d*) accessibility and connectivity.

7. Refer to Figure 7.16.

(*a*) Which countries load a greater weight of goods on to ships than they unload from ships? Explain the reasons.
(*b*) Which countries load less than half the weight of goods on to ships than they unload from ships? Explain the reasons.
(*c*) Describe and explain the contrasts shown in Figure 7.16 between countries at different levels of economic development.

8. Discuss the factors that appear to have influenced the character of the import trade of the countries shown in Figure 7.17.

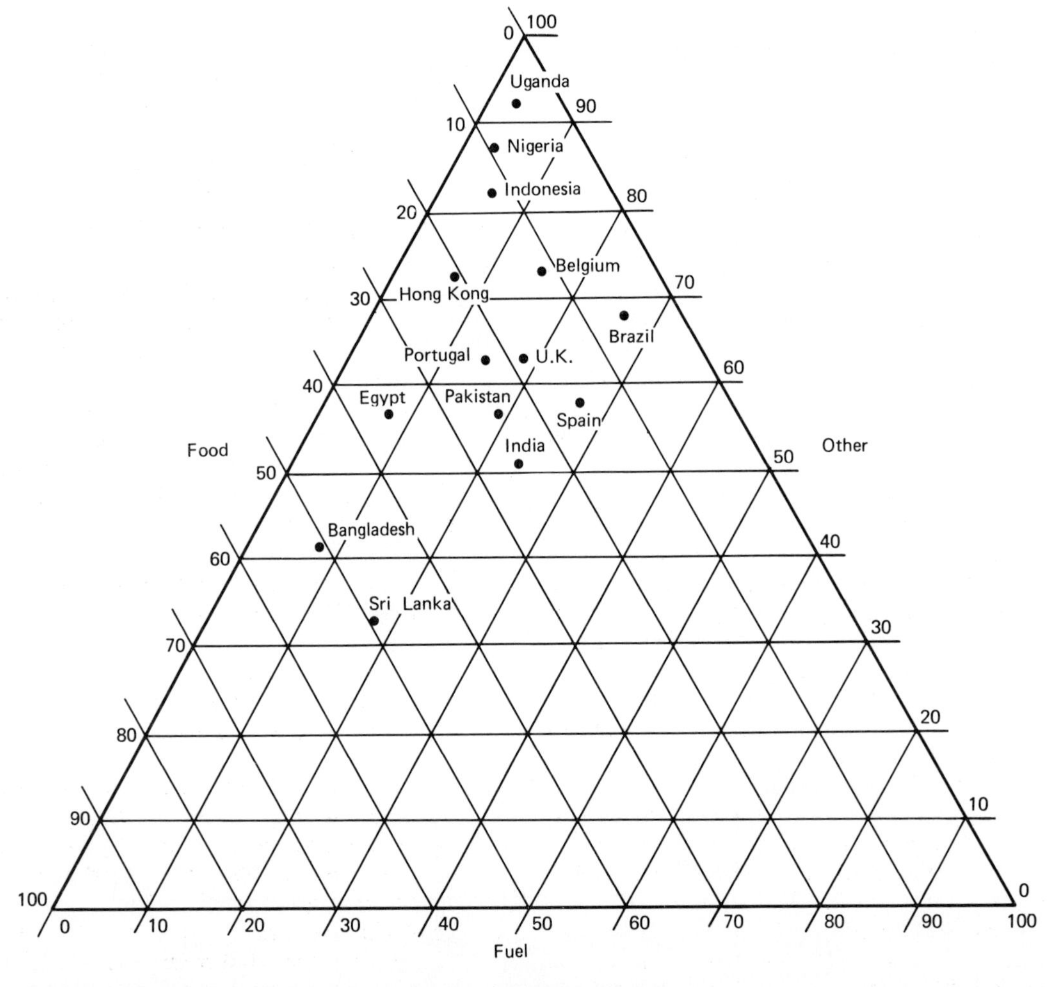

Fig. 7.17 Percentage shares of merchandise imports (by value)

8 Rural and urban settlement

8.1 Rural settlement patterns

NUCLEATED AND DISPERSED SETTLEMENT PATTERNS

A rural settlement pattern is made up of farm-houses and other buildings. Sometimes these buildings are widely scattered over the countryside and in other cases they are clustered together to form villages and hamlets.

Figure 8.1(*a*) shows an area of countryside in which the buildings are widely scattered. The area has a fairly even distribution of buildings; no building is very near to any other. This would be termed a dispersed settlement pattern. Such a pattern is frequently found in upland areas of Britain such as the Pennines and Wales.

In Figure 8.1(*b*), in contrast, the buildings form small groups (villages or hamlets). The buildings therefore are no longer scattered. Each building is very close to another building. Each village can be described as a 'nucleation' of buildings. Thus, Figure 8.1(*b*) illustrates a nucleated settlement pattern. Such a pattern is frequently found in rural areas of the English Midlands.

If we now take the village as the unit of settlement (instead of the individual building), it is clear from Figure 8.1(*b*) that, in the area shown, the villages are a considerable distance apart. This would be described as a fairly regular distribution of villages. In Figure 8.1(*c*), on the other hand, the 11 villages shown form three distinct groups, and no village is very far away from its neighbouring village. In such a settlement pattern the villages would be described as 'clustered'.

It is possible to calculate the extent to which villages are clustered (or single buildings are grouped into nucleations) by using *nearest-neighbour analysis*. The procedure is as follows:

(*a*) Measure the distance from each village to its nearest neighbour. Add together all these distances and then divide by the number of villages. This gives the average nearest-neighbour distance of the villages.

(*b*) Next we calculate what the average nearest-neighbour distance of these villages would be if they were distributed in a purely random (chance) pattern. This is done by using the following formula:

$$\text{average nearest-neighbour distance (random distribution)} = \frac{1}{0.5\sqrt{\left(\dfrac{\text{area of study area}}{\text{number of villages}}\right)}}$$

(*c*) The *nearest-neighbour statistic* is then found by dividing the measured average nearest neighbour distance [(*a*) above] by the average nearest-neighbour distance for the random distribution [(*b*) above].

(*d*) This will give a value between 0 and 2.15.

(i) If the value is 0 it means that the villages are all at the same location (and are

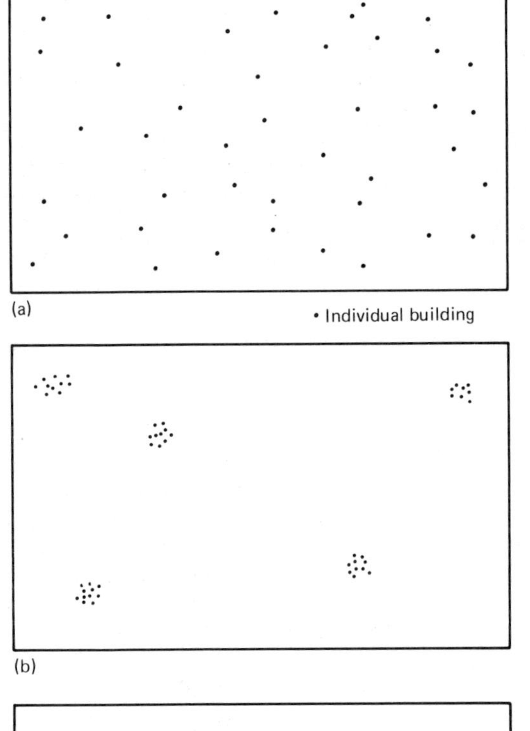

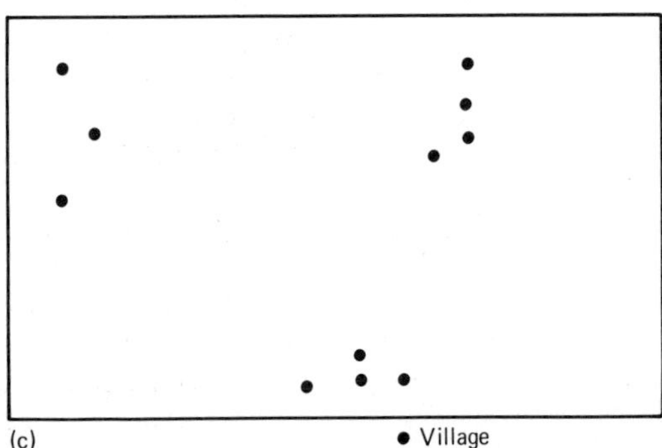

Fig. 8.1 Nucleated and dispersed settlement patterns

therefore no distance apart).

(ii) If the value is 1 it means that the villages are spaced as far apart as they would be if they were randomly distributed.

(iii) If the value is 2.15 it means that the villages are as far apart as it is possible for them to be in an area of this size. This means that they are arranged in a regular triangular pattern.

(iv) If the value is less than 1, it means that the pattern of villages may be described as 'clustered', and the smaller the value the greater the clustering.

(v) If the value is greater than 1, the pattern may be described as 'regular', and the greater the value the greater the degree of regularity.

(*e*) Figure 8.2 shows an area of land of 100 square kilometres in which there are ten villages in each case ((*a*), (*b*) and (*c*)). As the villages become further apart (a more regular distribution), the nearest-neighbour statistic changes from 0.63 (in (*a*)) to 1.27 (in (*b*)) to 1.9 (in (*c*)). The distribution shown in (*a*) is the most clustered, and that in (*c*) is the most regular.

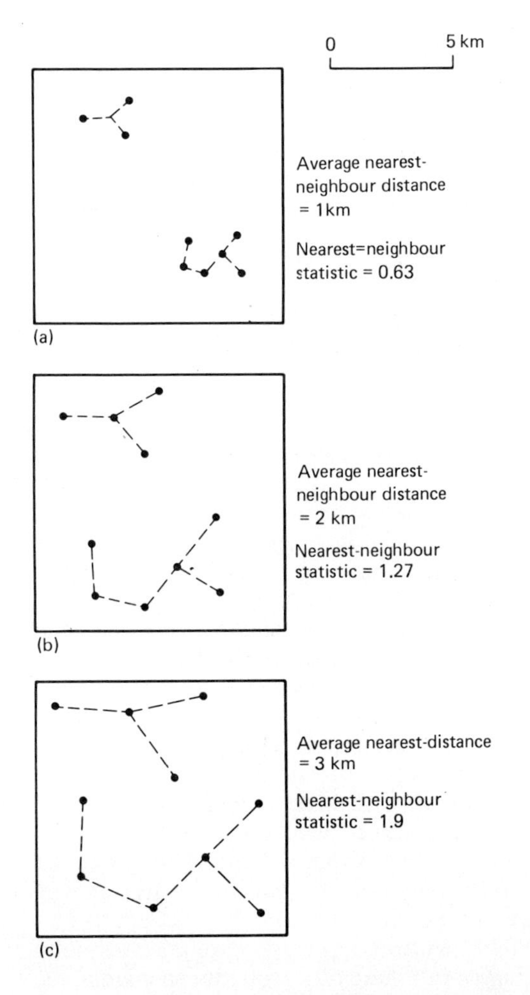

Fig. 8.2 Clustered and regular settlement patterns

CHARACTERISTICS OF NUCLEATED VILLAGES

The form (morphology) of the village

Nucleated villages differ greatly in general shape and also in the closeness of the grouping of individual buildings.

Some villages occupy a patch of land favourable for building, flanked by land on which building would have been difficult or undesirable. Thus, the shape of the villages closely reflects the shape of the patch of favourable land. This often occurs in hilly areas, where the villages extend in narrow strips along the valley floors, flanked by steeply rising land. Such villages are common in the Lake District, the Pennines and along the deep, narrow valleys of the South Wales coalfield. Figure 8.3(*a*) illustrates the principle. In the lowland areas of Britain, the shape of the village may be influenced by the pattern of land that is liable to flooding. Thus the village shape closely reflects the boundary of the land that is free from the danger of flooding. Thus, long, narrow villages may occupy a river terrace above a flood plain and more compact villages may occupy a dry 'island' site in a flood plain (Fig. 8.3(*b*)).

The need for defence in past times may influence village shape. In some villages the houses are grouped around a central open space. These are often called 'green villages' (Fig. 8.3(*c*)). Particularly in Mediterranean countries, villages often occupy hillside or hilltop sites and their shape reflects the shape of the hill. Villages of compact shape may have been set up within the loops of river meanders.

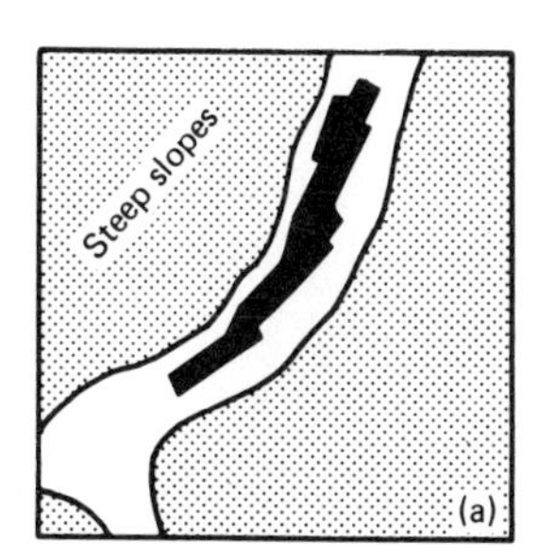

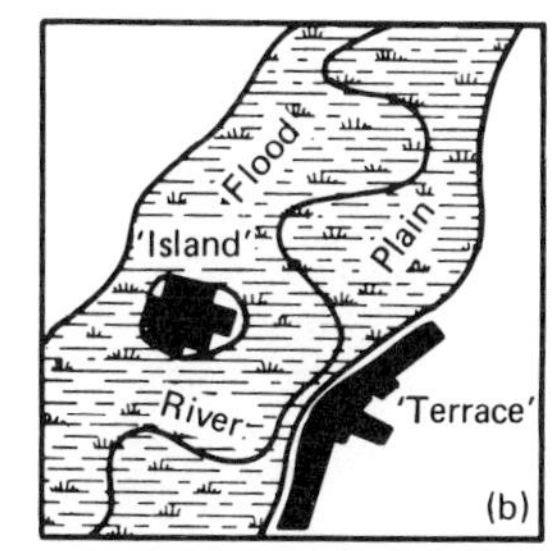

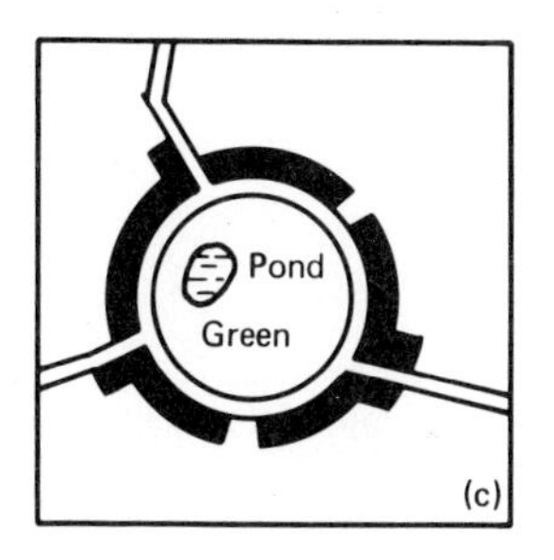

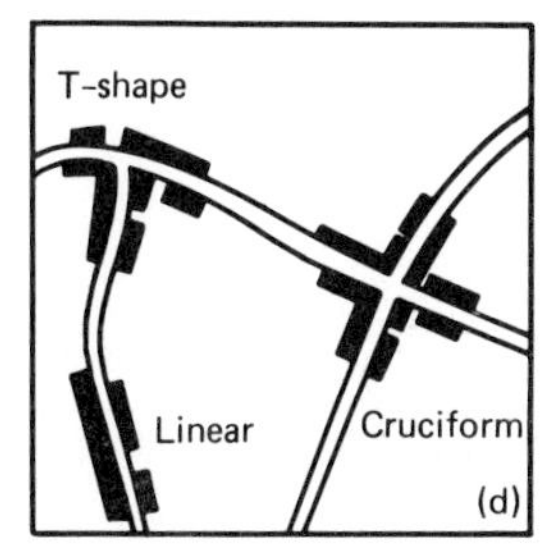

Fig. 8.3 The morphology (form) of villages

In central Wales, some villages are clustered around an ancient church established by a Celtic missionary, and the churchyard forms an open space in the centre of the village.

Many villages owe their shape to the pattern of the road network. They are sometimes long and narrow, with buildings along each side of a road. Or they may be T-shaped or cross-shaped at road junctions (Fig. 8.3(*d*)).

Sometimes the separate buildings of a village may be fairly far apart. This is the case with crofting settlements in northern Scotland. Although they may be described as 'nucleated' in relation to the large areas of unpopulated land which separate them, the buildings within them are often fairly dispersed (anything up to 100 metres apart). Each croft often has its own patch (up to about two hectares) of arable land.

The distribution of nucleated villages

Groups of nucleated villages in many parts of England form characteristic patterns of distribution. In this case, we are regarding the village, rather than the single building, as the unit of settlement.

Spring line villages are located along the base of the scarp slope or the dip slope of a chalk cuesta, where springs issue from the water table. They are often situated one to two kilometres apart. Parish boundaries are arranged so as to give each parish a share of the dry chalk land and the better watered clay land (Fig. 8.4(*a*)). Spring line villages are also found flanking Jurassic limestone and Greensand cuestas.

Dry valley villages are found in dry valleys in a chalk cuesta, where a water supply can be obtained by the use of shallow wells reaching down to the water table (Fig. 8.4(*a*)).

Dry point villages tend to form 'lines' along the edge of areas of marshland, or either a double or single line along the edge of a river's flood plain (Fig. 8.4(*b*)).

In upland areas, lines of villages tend to follow the major valleys. If a valley runs from east to west, the villages are usually located on its north side to avoid being in the shadow of the high southern valley side (Fig. 8.4(*c*)).

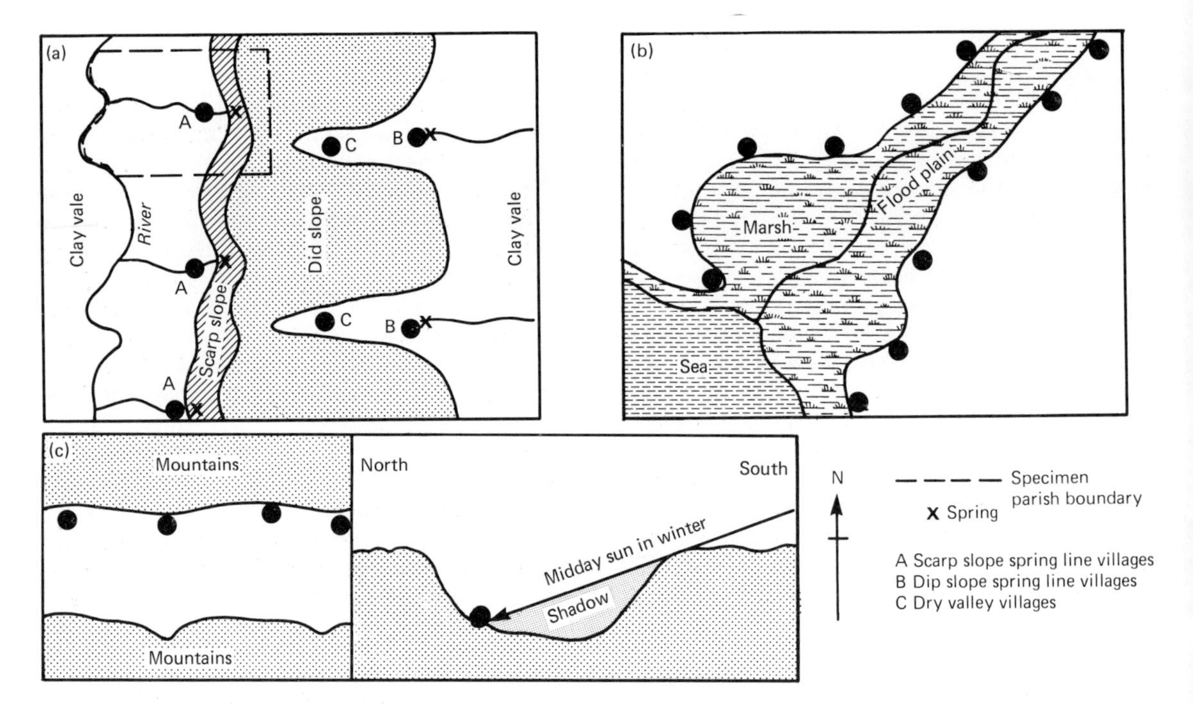

Fig. 8.4 Patterns of nucleated villages

FACTORS FAVOURING THE NUCLEATION OR DISPERSAL OF RURAL SETTLEMENTS

If rural settlements were entirely composed of the homes of farmers who were engaged in farming their own areas of land, it seems that a dispersed settlement pattern would be much more efficient than a nucleated one. In Figure 8.5 it is clear that if the farmhouses are clustered together near the centre of the whole area of farmland each farmer has to travel a greater distance to work than if the farmhouses are scattered among the various holdings. There may however be other considerations which encourage nucleation of settlement.

Water supply

During the Anglo-Saxon colonization of England after Roman times each settlement needed a

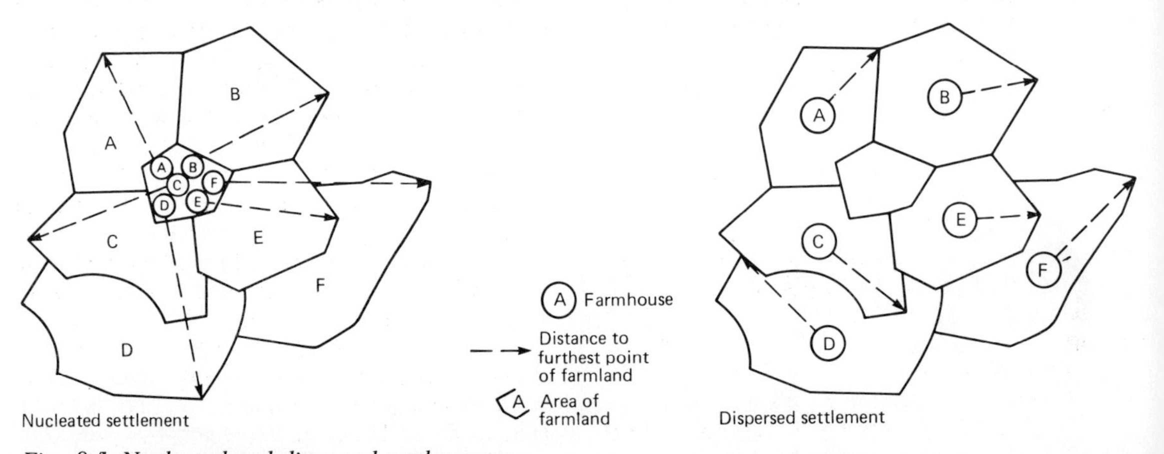

Fig. 8.5 Nucleated and dispersed settlements

supply of water above all else. Water is used very frequently and is bulky and heavy to carry. Hence, nucleated settlements were founded near supplies of water, as at a spring line or on the banks of a river. In modern times it is less necessary for settlements to be located at a supply of water. Water can now be piped to individual buildings, and so can be easily supplied to dispersed settlements.

Law and order
At different times in various parts of the world it has been an advantage for settlements to nucleate for reasons of defence. In Norfolk, villages grew around medieval moated manor houses. The hill villages of Italy were established for defence reasons, as were the very small 'castle villages' of northern Scotland, such as Inveraray. With the establishment of law and order, there is less reason for settlements to nucleate, so dispersal tends to take place.

The land tenure and inheritance system
In the Anglo-Saxon open field system each peasant was allocated a number of scattered strips of land in the open fields. Hence a home in the nucleated village would be fairly centrally placed in relation to these strips.

In Wales the Anglo-Saxons had little influence in medieval times. A system of inheritance of land known as 'gavelkind' was practised whereby, at the father's death, all his land was divided between all his sons. This led to dispersal of settlement as the sons set up homes on their plots of land.

Also, in upland parts of Wales and northern Britain a dispersed settlement pattern exists, with farmhouses standing upon relatively small patches of cultivable land.

Population change
An increase in population could lead to overcrowding in a nucleated village. Hence, people might move away from the village and set up farms on the wasteland beyond the boundaries of the open fields. In some cases they might occupy clearings in the forest which previously had only been used during summer.

The enclosure of land
In the late eighteenth and early nineteenth centuries much of the land of Britain was divided into enclosed fields by a series of Enclosure Acts of Parliament. At this time the old open fields were enclosed and enclosed fields were created in fenland, moorland and downland which previously was of little value. New farms were set up on this enclosed land, thus creating a dispersed settlement pattern.

In the Pennines much moorland was walled into fields at this time and dispersed farms were set up on this newly enclosed land. Parliamentary enclosure created large, generally rectangular fields and straight roads. These now contrast greatly with the smaller, irregularly shaped fields, and the winding roads, on the lower land which had been enclosed much earlier.

When new land was opened up to settlers in the nineteenth century in the USA and Canada, the land was divided into rectangular sections of one square mile which contained four family farms. Thus, in general, a dispersed settlement pattern, comprising four farms per square mile was created.

THE ANGLO-SAXON SETTLEMENT OF EASTERN AND SOUTHERN ENGLAND

Much of the pattern of rural settlement that now exists in eastern and southern England was first laid out in the centuries following Roman times which are generally known as the 'Dark Ages'. During these centuries Anglo-Saxon invaders advanced into England and set up permanent settlements. The Anglo-Saxon settlement pattern was strongly influenced by the physical environment.

The Anglo-Saxons entered from the east and south-east, usually along the rivers and, at an early date, established settlements on gravel terraces near major rivers such as the Thames, and also along the Sussex coast, where the soils were rich and loamy. They established villages which can now be identified by the -ing or -ings suffix of their place names. In Old English such place names referred to groups of people rather than actual places. Other place name suffixes which represent very early Anglo-Saxon settlements in south-east England are -ham and -ton.

At this early stage, the Anglo-Saxons tended to avoid areas of less fertile soils such as the sandy or gravelly areas of the central Weald and the New Forest. They also avoided areas with heavy clay soils which at that time would be heavily forested.

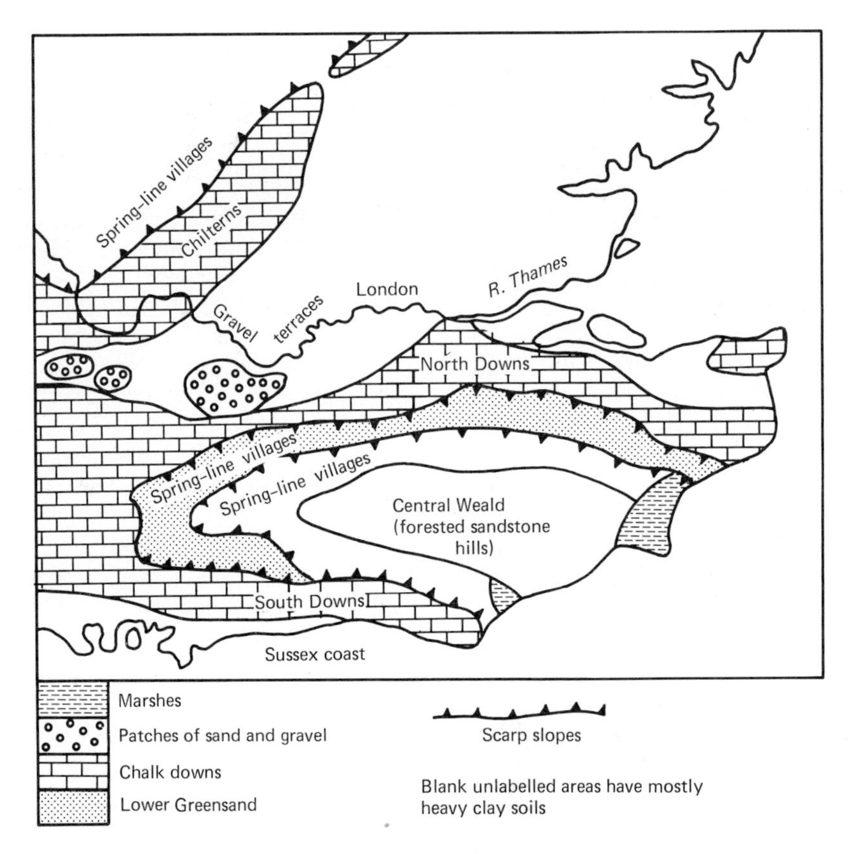

Fig. 8.6 Physical influences on the pattern of Anglo-Saxon settlement in South East England

As they advanced inland, the Anglo-Saxons set up a great many spring-line villages at the foot of the chalk downlands. These locations had many advantages, including:

(*a*) an assured, good-quality water supply from the springs;
(*b*) an area of rich cultivable soil at the foot of the chalk scarp where downwash from the chalk had mixed with and lightened the clay soil;
(*c*) areas of grazing land on the open chalk downlands and a supply of timber from the clay lands nearby.

Thus Anglo-Saxon villages were set up beneath the scarps of the chalk downs surrounding the Weald, and along the chalk scarp which includes the Chilterns and the Wiltshire downs. The Greensand ridges also provided settlement sites.

Later, settlement began to extend into the areas which had earlier been avoided, especially in the clay lands where the process of removing the natural forest began. The establishment of new, secondary settlements in forested areas of the Weald and elsewhere is indicated by place name suffixes such as -ley, -den and -field.

The general physical conditions that influenced the pattern of Anglo-Saxon settlement in south-east England are shown in Figure 8.6.

8.2 World patterns of towns and cities

URBAN POPULATION

The proportion of the population living in towns and cities varies greatly over the world. In some countries as few as 5 % of the population lives in large towns; in others the figure is as much as 60 %. The level of urbanization of any part of the world depends mainly upon the stage it has reached in its economic development (pages 36–41).

Traditional rural societies

In traditional rural societies a large proportion of the population is engaged in agriculture. Usually settlements are small and dispersed, so that farmers live as near as possible to the fields that they cultivate, thus saving the time and effort of travel to work. Similarly, communities that depend on hunting or fishing rarely live in towns.

Also, in many such areas, there is a subsistence economy, with very little farm produce available for sale. Hence towns could not be supported by produce from the countryside. Such rural populations produce their own food locally and also, by means of domestic handicrafts, provide the simple manufactured goods that they need.

Usually only small towns will exist in such areas, often for the purpose of organizing the small amount of local trade which is carried on. These towns may have periodic markets, held every week or every ten days which are visited by travelling salesmen.

In some cases, a single sizeable city may exist which may be an administrative centre, a military centre or a coastal trading port. However, such cities are the result of some external influence, often of European origin, which has affected the traditional society. It will not be a 'natural' growth out of the local subsistence economy.

Most developing countries have a low level of urbanization such as this. In the Philippines, for example, less than 15% of the population lives in large towns with populations of 100 000 or over. In Nigeria the figure is about 12%. Some countries, such as Tanzania, have only one large town, even though their total population is well over 10 million.

Advanced urbanized societies

As soon as non-primary (e.g. non-agricultural) economic activities, such as manufacturing industry and the provision of services, begin to develop, the level of urbanization begins to increase.

Improvements in transport facilities make it possible for farmers to specialize in the production of particular products. They then market their surpluses in urban centres. Also, the mechanization of farming leads to a higher output per farm worker, thus producing a surplus of farm produce which can support non-agricultural urban populations. As commercial farming replaces subsistence farming (pages 58–59), towns and cities are able to develop.

As manufacturing industry develops on a larger scale, it tends to become concentrated in towns where it can obtain the benefits of proximity to other, related, industries (external economies) and also internal economies of large-scale production. Thus, many industries can be carried on more efficiently in towns.

Also, in these growing towns, service (tertiary) industries (page 39) develop, such as the retail and wholesale trades, business organization, education, entertainment and tourism.

Thus, as economic development takes place, the level of urbanization increases. Eventually, though, it appears that urban development begins to slow down. As cities grow in size many problems begin to appear and population begins to move outwards into the countryside. Thus, in some advanced countries such as the USA and Britain much suburban development of houses and industries has taken place on the rural fringes of the large cities.

The world pattern of urbanization

Well over 2000 years ago towns and cities were created by the most advanced peoples of the time in the Nile, Euphrates and Indus valleys and in China, but rapid urbanization really began with the Industrial Revolution in Europe, particularly in the nineteenth century. This was associated with the great improvements in transport that occurred at this time. Urbanization then developed in areas such as North and South America and Australasia, to which Europeans migrated, and then in the Far East to which European ideas spread. In the late twentieth century, rapid urbanization is taking place in the developing countries as population migrates from the countryside to the cities (pages 51–53).

In most of the countries of north-west Europe over 30% of the population lives in large towns of at least 100 000, but this figure decreases slightly to the east and south where economic development is not so advanced. Bulgaria and Italy, for example have relatively low percentages.

In general, over the world, the higher the average living standard (GDP per capita) the greater the proportion of people living in large towns. This is illustrated by a sample of 12 countries in Figure 8.7.

High percentages of the population live in large towns in Europeanized countries such as Canada, Argentina and Australia, all with at least 50%. A

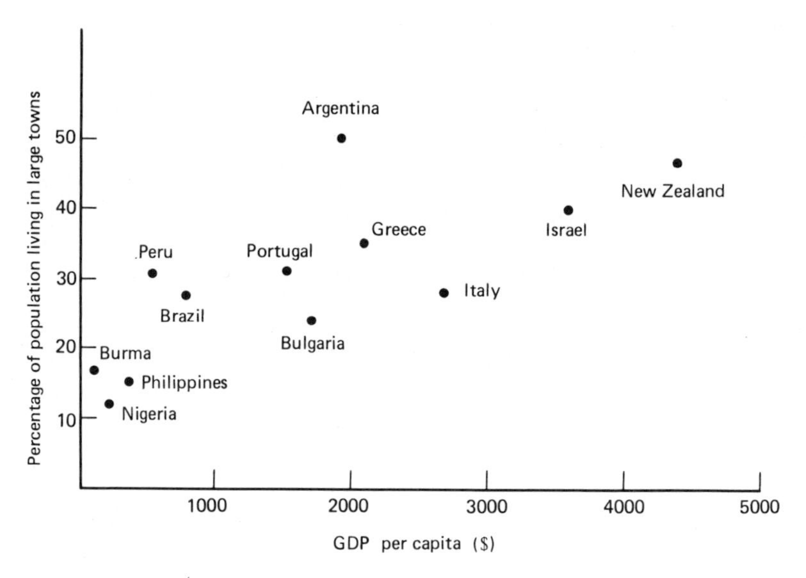

Fig. 8.7 Urbanization in relation to living standard

feature of some of these Europeanized countries is the very large percentage of their population that lives in the largest city of all (about 40% in Uruguay and over 30% in both Argentina and Chile). This is a condition known as 'primacy'.

THE WORLD DISTRIBUTION OF 'MILLION CITIES'

In the twentieth century the very large city with over one million inhabitants has appeared.

The growth of 'million cities'

Since the 1920s 'million cities' have been created at a very rapid rate. Their number has increased from about 20 in the 1920s to about 80 in the 1950s, 110 in the 1960s and 160 in the 1970s. During this period the 'five million city' has made its appearance. There were six of these in the 1950s, 10 in the 1960s and 16 in the 1970s. During the 1970s cities have grown even larger and there are at least four '10 million cities'.

The distribution of 'million cities'

In general, since the 1950s, Europe has had between 25 and 30% of the world's 'million cities', Asia has had 30–40%; and North America has had 15–20%. Relatively few have grown in Africa and Australasia. These percentages have remained remarkably constant as the number of 'million cities' has increased.

The 'five million cities' have had a similar distribution, concentrated mostly in Europe, Asia and North America but, in the 1970s, Latin America has seen a considerable increase, Sao Paulo, Rio de Janeiro and Mexico City being added to Buenos Aires which was one of the earliest cities to reach this size. Also, in the 1970s, Calcutta and Bombay have been added to Tokyo Shanghai and Peking which had already reached this size in Asia.

The earliest '10 million cities' were Shanghai and Tokyo in Asia, Mexico City in Latin America and New York.

As the number of 'million cities' has increased their distribution has tended to become more and more concentrated in lower latitudes nearer to the equator. The earlier 'million cities' tended to be further from the equator than the more recent ones. The 'average latitude' of 'million cities' has moved towards the equator at a rate of about 2° in every 10 years.

The changing distribution of 'million cities' is shown in Figure 8.8. It closely reflects the different rates of economic development in different parts of the world.

In the 1950s clusters of 'million cities' existed in Europe (particularly Great Britain), north-eastern USA, and the Far East (particularly Japan, China and India). Elsewhere they were more scattered, as

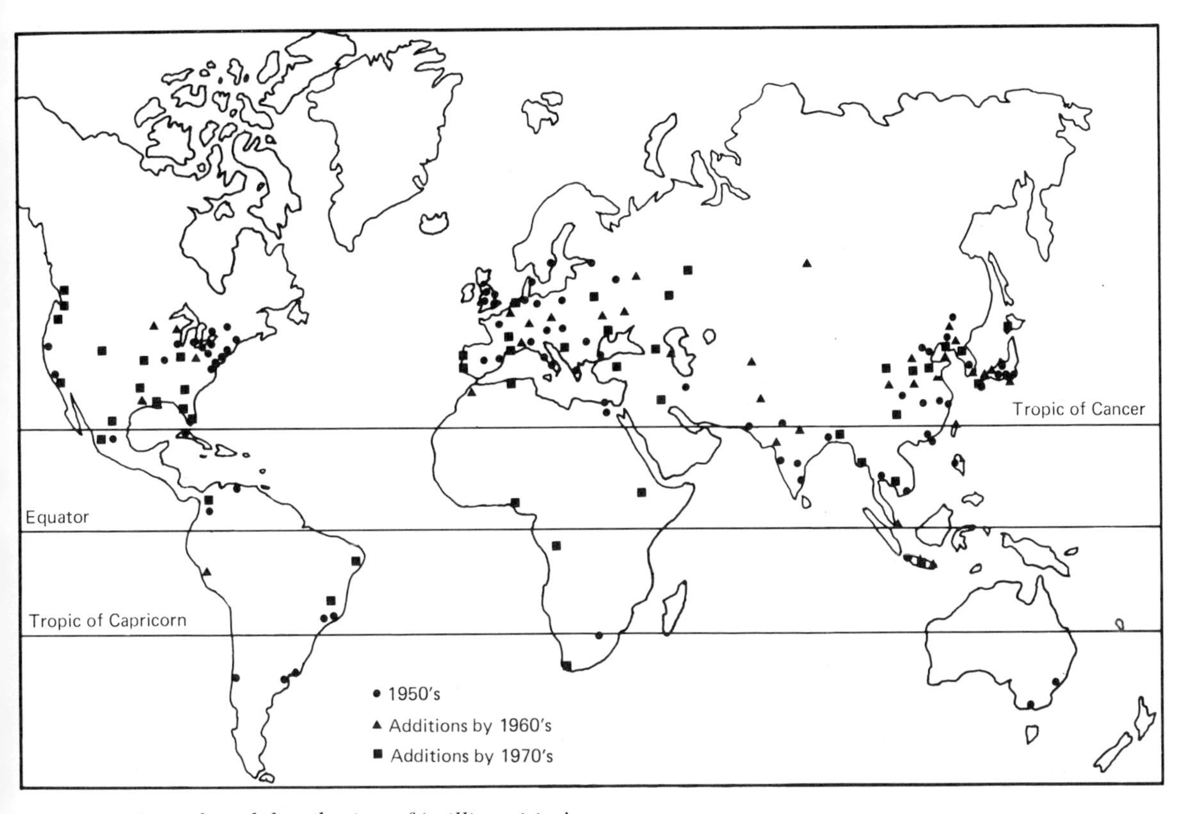

Fig. 8.8 Growth and distribution of 'million cities'

along the west coast of the USA, and the coasts of South America, south-east Asia and Australia. Africa had only three, one in Europeanized South Africa (Johannesburg) and two (Cairo and Alexandria) in Egypt.

In the 1960s new 'million cities' came into existence mostly on the fringes of existing clusters, as in the Great Lakes area of the USA, in European USSR, northern China (Manchuria) and Japan. Only Lima (Peru) was added in Latin America, and Casablanca in Africa.

In the 1970s there has been a considerable change in the distribution and new 'million cities' have appeared a considerable distance from the earlier clusters. Several have appeared for example in southern and central USA (Miami, Dallas, Atlanta, New Orleans and Tampa in the south, and Denver and Kansas City in the centre). Also a small cluster (Vancouver, Seattle and Portland) has developed on the Pacific coast. The other major development has been the rise of the 'million city' in Africa in which Algiers, Lagos, Addis Ababa, Cape Town and Kinshasa (Zaire) have reached this size. 'Million cities' also have developed to the east and south-east of the main urbanized part of the USSR.

A feature of the general pattern of 'million cities' shown in Figure 8.8 is the tendency for many countries to have just one of these cities, as in Europe (Sweden, Denmark, Poland, Austria, Hungary, Romania, Greece), South America (Venezuela, Peru, Chile, Uruguay, Argentina), Africa (Algeria, Morocco, Nigeria, Zaire, Ethiopia), the Middle East (Iraq, Iran) and south-east Asia (Bangladesh, Burma, Cambodia, Thailand, Vietnam).

CITY-SIZE DISTRIBUTIONS

The sizes (population totals) of cities within individual countries also tend to form certain patterns. Studies of the population totals of the cities of various countries have suggested a tendency for the sizes of the leading cities to be related to the size of the largest city of all in the country.

The so-called 'rank-size rule' is expressed in the

form of a simple equation $P_r = P_1/r$. This means that the population (P) of a city of any rank (r) tends to be equal to the population of the country's largest city (P_1) divided by the rank (r) of the city that is being considered.

The rank-size rule

By the rank-size rule, if a country's largest city has a population of 10 million, we would expect the second largest to have 5 million (10/2), the third largest to have about 3.3 million (10/3), the fourth largest to have 2.5 million (10/4), and so on.

Represented on a graph where population is plotted against rank, this gives a curving line (Fig. 8.9(a)). If, however, logarithmic scales are used on the axes of the graph the line becomes straight, at an angle of 45° (Fig. 8.9.(*b*)). It is much easier to compare any country's city-size distribution with this straight line than with the curve shown in Figure 8.9(*a*).

Similarly, in Figure 8.9(*b*), rank-size distributions are shown on the graph for largest cities of six million and three million. This is the simplest form of the rank-size rule and is the one most commonly used, but straight lines sloping at

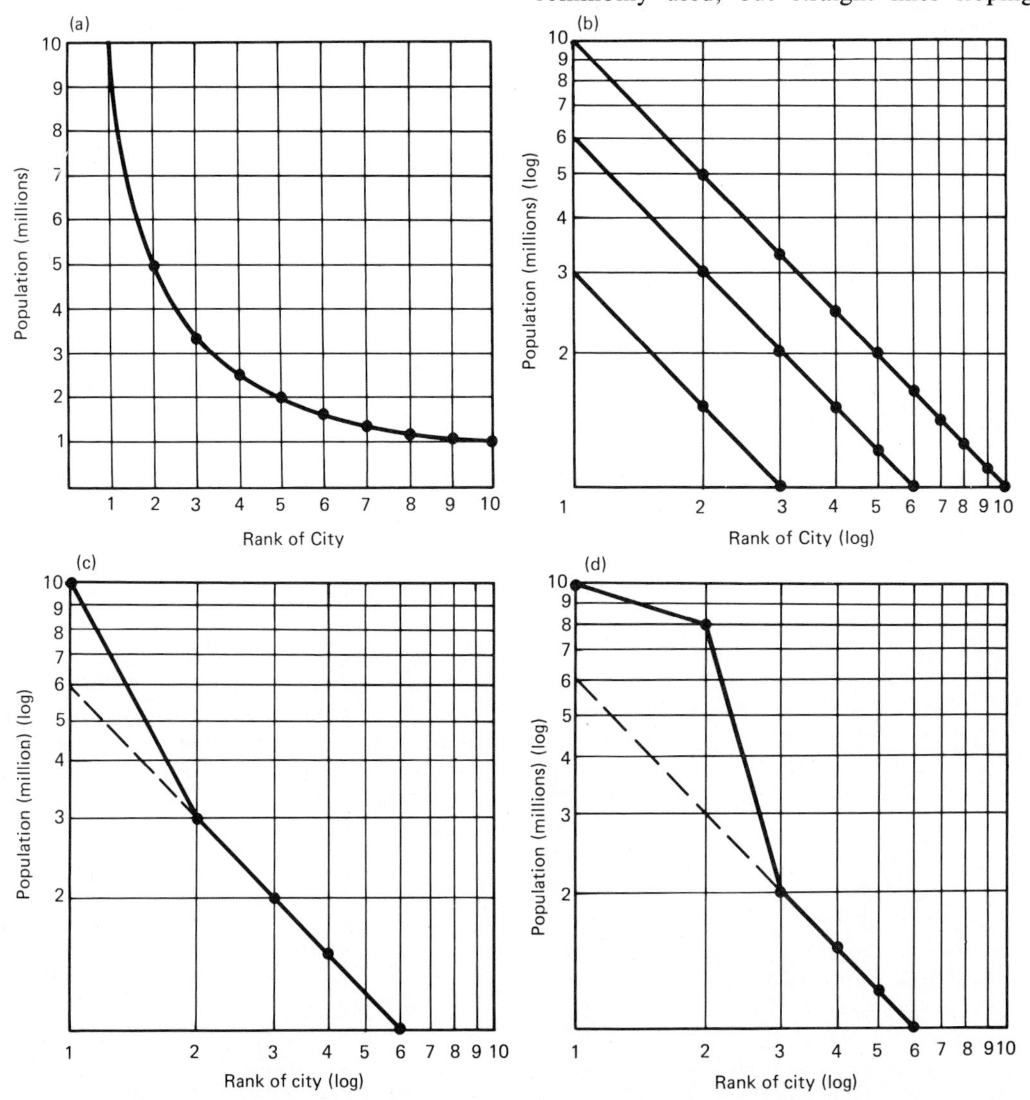

Fig. 8.9 The rank-size rule

different angles would also show rank-size relationships.

In practice it is quite rare to find a city-size distribution that is similar to the rank-size rule. Certain common variations tend to occur. One of these is the 'primate' distribution in which the largest city is much larger than it should be according to the rank-size rule. This is shown in Figure 8.9(*c*). In this graph, if we project the line of the cities ranked second to sixth we find that the 'rank-size' population for the largest city is six million. In fact the population of the largest city is 10 million. Sometimes there may be two 'primate' cities as in Figure 8.9(*d*). This is known as a 'binary' distribution.

Factors influencing city-size distributions

Little is known about the factors that cause the appearance of rank-size distributions. It is said that countries that are large, have been urbanized for a long period of time, and have developed highly complex economic and political organizations tend to have rank-size city distributions. On the other hand, small, somewhat primitive countries, which until recently were colonies of European countries are said to be likely to have primate distributions. In practice many exceptions to this rule can be found.

One great problem in interpreting a city-size distribution is the fact that the country's area may have changed through time. For example, if a number of colonies were originally set up along a coastline, a primate distribution might evolve in each one, especially if much of the economic and political activity were to be concentrated at the port city. At a later date, these colonies might be combined into a single country. Hence all the former primate cities would now be in the same country, so it is unlikely that the new country would have a primate city distribution. This applies in Australia for example, where Sydney and Melbourne are both about the same size.

Some countries in Europe, such as Italy and West Germany have only recently been created by merging a number of separate states. These are unlikely to have primate distributions. Conversely, it is possible for a primate distribution to occur in a highly developed advanced country. Examples are Austria and Denmark. This is because their largest cities were, until recently, the centres of a much larger region than the present country.

Examples of city-size distributions

In the 1970s it appears that many countries have primate city-size distributions. In the following countries the largest city is at least 10 times larger than the second city: Argentina (Buenos Aires), Chile (Santiago), Peru (Lima), Uruguay (Montevideo), all in South America, and Hungary (Budapest) in Europe. In many other countries the largest city is at least five times larger than the second city, as in Austria, France, Iran, Iraq and

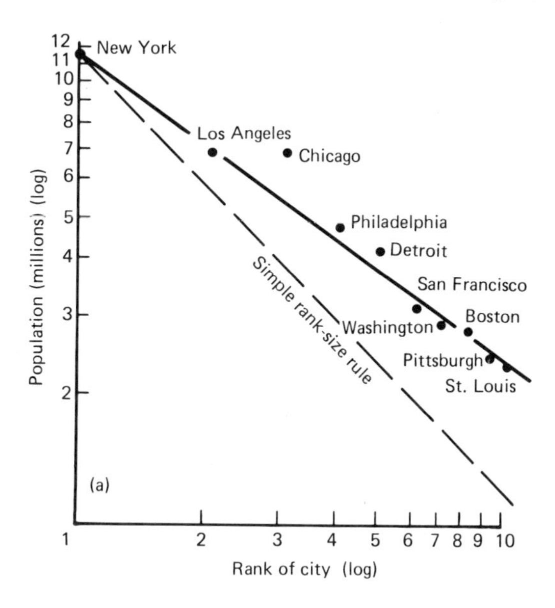

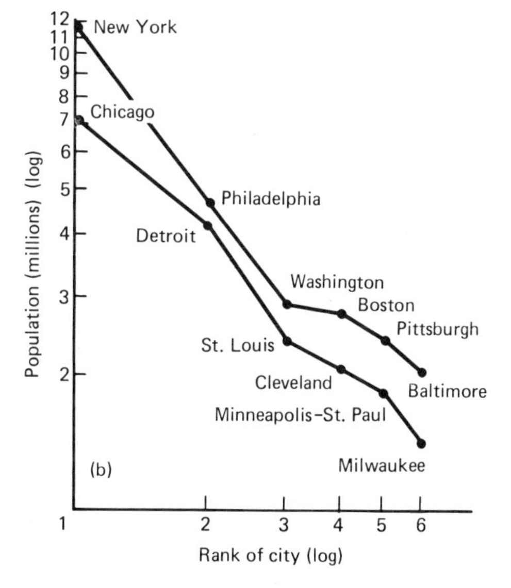

Fig. 8.10 City-size distributions in the USA

Mexico, for example. Not all of these could be classified as developing countries. France in particular has had a long history of town growth, yet it has not developed a rank-size distribution.

Some countries tend to have binary distributions where two cities are very much larger than the remainder. Examples are Brazil (Rio de Janeiro and Sao Paulo), Pakistan (Karachi and Lahore) and Spain (Madrid and Barcelona). These could be associated with a division of functions between two cities, a port on the coast and an industrial or administrative centre inland.

The USA has an approximate rank-size distribution as shown in Figure 8.10(*a*), although the cities decrease in size rather more slowly than for the simple version of the rank-size rule. Nevertheless, the points on the graph form almost a straight line except for Chicago, which is rather too large.

If rank-size graphs are drawn for the cities of eastern and central USA separately (Fig. 8.10(*b*)) in each case there is almost a rank-size relationship. But the steepening of the lines at Washington and St. Louis suggests that New York, Philadelphia, Chicago and Detroit might be classified as primate within their own particular regions of the USA.

URBAN FORM

The 'form' or 'physical morphology' of a town is the types of buildings, streets and roads and the patterns that they make within the town. The buildings and street plan of a town may last for a great many years, even though their *functions* change. In many towns there are buildings at least 500 years old, and parts of the street plan may be even older. Thus the form of a town is to a great extent a summary of the town's history.

The form of towns in advanced countries

Many British towns have been in existence for many centuries and, during this time they have added a variety of elements to their structure. It is usually possible to distinguish three broad zones of contrasting urban form: the central area, the inner zone and the outer zone. Figure 8.11(*a*) illustrates these three zones for Brighton.

In many cases the central area was created in medieval times when streets were used by pedestrians rather than vehicles. Streets are often narrow and winding. In some cases, as at Chester and Gloucester, the street plan was first laid out in Roman times, so the streets form a rectangular pattern with major streets leading to the gates in the old town walls.

In modern times, when the central area has become a central business district, old buildings may have been cleared to make way for shops, offices and hotels, and new wider roads may have been made to allow traffic to flow more easily. In some cases, as at Chester, new shopping developments have been located to the rear of the buildings which line the ancient streets. The

The pattern of urban growth in Bolton

central area is almost fully built up. There are few open spaces apart from car parks and public gardens and possibly the grounds of a castle or a cathedral. New buildings, such as office blocks, department stores and multistorey car parks are built high. The central area usually has the highest buildings in the town.

The railway station is usually located at the edge of the central area. At the time railways began to appear in the mid-nineteenth century most towns consisted only of what is now the central area. The railway station was built on the outskirts of this old town so as to avoid unnecessary demolition of buildings.

During the late nineteenth century many British towns grew rapidly as people migrated into them from the countryside. In this period a fine-grained pattern of rows of terraced houses with closely spaced parallel streets (the inner zone) was (Fig. 8.11(*b*)) created, often with a scattering of industrial buildings. In modern times these buildings have tended to deteriorate, particularly near the fringe of the central business district. Hence 'urban renewal' has taken place. The old terraced houses and industrial buildings have been demolished and replaced by high-rise residential flats. More recently, high-rise flats have become unpopular and planned estates of small houses with gardens linked by traffic-free pedestrian walkways have been built instead.

In the twentieth century since about 1920, many British towns have extended far into the surrounding countryside (Fig. 8.11(*a*)). Large estates of semidetached houses and bungalows (outer zone) have been built, served by systems of planned, relatively widely spaced avenues. These suburban estates have also developed around old village centres and have joined them to the main urbanized area. Large industrial buildings have also been set up in these suburban areas for industries that demand large amounts of space. All this has been possible because of the development of

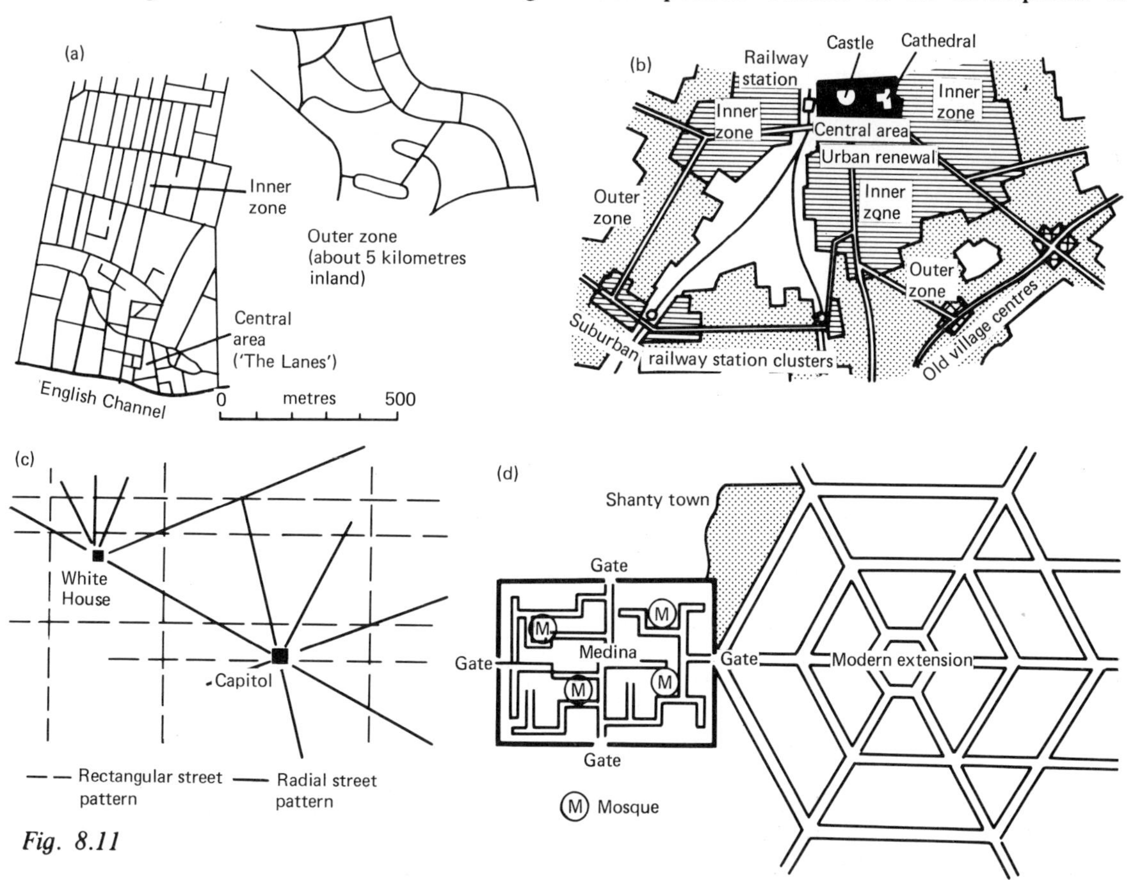

Fig. 8.11

personal transport by the family motor car and lorry transport for industrial goods. Such suburban areas are often served by wide, dual-carriageway roads intersecting at large, space-using roundabouts.

The general shape of British towns has often been strongly influenced by physical features. Low-lying flood plains and steep hillsides have usually been avoided. Also, open spaces may be left in the urban settlement pattern because they were designated as parks or cemeteries or commons. The road and railway pattern has also influenced town shape. Building has tended in the past to extend along major roads and to cluster around suburban railway stations. Some of these relationships are illustrated in Figure 8.11(*b*).

In North America, urban development began later than in Britain so the medieval core is absent. In some cities of eastern USA, such as New York, small areas with an irregular street plan may indicate the extent of the original port settlement, as at the southern tip of Manhattan Island, but most of the town has a geometrically regular, usually rectangular, street plan. Washington, a planned capital city, has a complicated pattern of streets radiating from both the Capitol and the White House, upon which is superimposed a north-south and east-west rectangular pattern (Fig. 8.11(*c*)). Like Washington, most of the towns of central USA were built from the beginning on virgin land. Hence, the whole street plan is usually rectangular, even in the central area.

Most of the cities of the USA also have very large areas of twentieth century 'suburban sprawl'. The level of car ownership in the USA has always been higher than in Britain. Hence, the population has generally been more mobile and able, for example, to travel greater distances to work. Americans have also tended to demand larger plots of land for their houses. Los Angeles has been described as the 'largest' city in the world, measured in terms of the space that it covers.

In many European cities, twentieth century 'urban sprawl' is less noticeable. In general, flat dwelling in more common than in Britain and there are fewer low-density housing estates on the outskirts of towns.

The form of towns in developing countries

Cities in developing countries also illustrate the history of the area in which they exist.

In the Middle East, along the North African coast and to the east, there is often an ancient walled city (medina) which contains mosques and workshop and market areas (suqs) (Fig. 8.11(*d*)). The narrow streets of the medina form a complex pattern with many cul-de-sacs. This part of the city was created at the time of Arab domination before the tenth century. Houses in the medina have few windows facing the street, but often have an off-street courtyard. The modern part of the city usually has a European appearance, with straight, wide streets forming geometrical patterns. Shantytowns may exist near the outskirts (Fig. 8.11(*d*)).

The form of some of India's cities also illustrates their history. Delhi has a nucleus with narrow winding streets near a fort (the Red Fort) built to protect the area from invaders from the north-west. During the British period of influence a modern capital city (New Delhi) was created well away from the ancient city, with wide avenues forming a street plan composed of triangles and hexagons.

In south-east Asia there were few cities before the Europeans arrived. The European city was usually established close to the harbour. Its form usually reflects the nationality of the Europeans who created it. Jakarta (Indonesia), for example, established by the Dutch as Batavia, came to have a pattern of narrow streets, with tall buildings and canals, rather like Amsterdam. Ho Chi Minh City (Saigon) in Vietnam was set up by the French and has wide, tree-lined boulevards like Paris. Most south-east Asian cities have also acquired a Chinese quarter, as the Chinese have immigrated, with closely packed shops and houses along narrow streets. They also have shantytowns with very irregular settlement patterns.

URBAN FUNCTIONS

General land use patterns in towns

Land uses which are mostly concentrated in towns derive benefit from being close to one another (agglomeration). Some of them benefit greatly from a location near the centre of the urban area, which has a high level of accessibility to the rest of the town.

Consider a town that is circular in shape and within which it is equally easy to move in all directions. The geometrical centre of this town will be its most accessible point; it is as near as possible to all parts of the town. Suppose that shops, industry and residence are competing for space in this town and that each part of the town's area will be taken over by the land use that makes the highest bid for it.

Shops need to be accessible to their customers, since people must travel to shops before the shops can earn any profits. Most shops also can carry on their business in a very small amount of space, which is used very intensively. Hence shops are able to bid a high price for the use of land in the most accessible (i.e. the central) part of the town (Fig. 8.12). Moving away from the town centre, however, the land rapidly becomes less attractive to shops. Hence the bid-price curve for shops slopes downwards very steeply away from the town centre (Fig. 8.12).

Industry also places a value on accessibility within the town. It needs access to its workers who may live in various parts of the town and it may market its products throughout the town. Some industries are able to use space very intensively but, in general, industry uses space less intensively

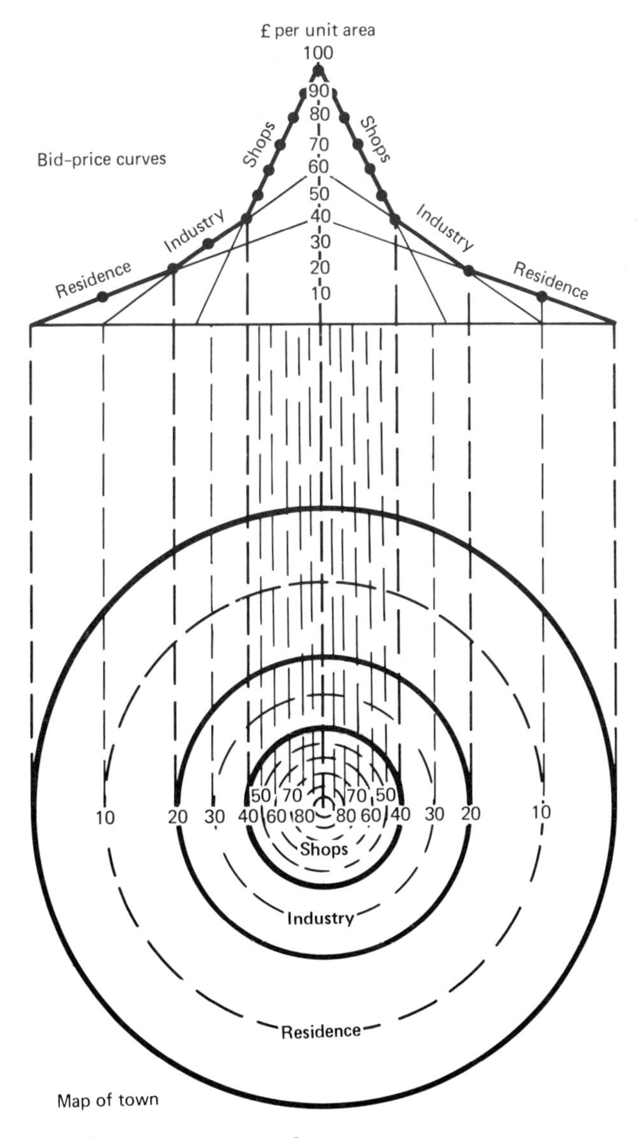

Fig. 8.12 Competition for space in a town

than shops. Industry therefore may prefer a central location but its need for one may not be so great as with shops. Hence, industry's bid for land at the town centre is lower than that of shops, but, moving away from the town centre, it declines less quickly (Fig. 8.12).

The bid-price curve for residence is flattest of all. If these three bid-price curves are drawn on the same graph it can be seen that shops outbid both industry and residence for a town centre location. Industry outbids both shops and residence a short distance away from the town centre and, on the map, forms a ring around the central concentration of shops. Residence outbids the other two functions in the rest of the town. Thus we would expect a sequence from shops to industry and residence from a town centre to the outskirts (Fig. 8.12).

The pattern of land values in this imaginary town reflects the bids made for land by these three land uses. It rises to a peak in the shop zone in the town centre, but declines very steeply immediately away from the centre. In the industrial and residential zones land values are much lower and they decline less steeply outwards.

Of course the land use pattern shown in Figure 8.12 could not exist in real life. Towns are never perfectly circular in shape and it is never possible to move equally easily in all directions within a town. In all towns movement is channelled along transport routes, especially roads.

Figure 8.13 shows another imaginary town, but this one has a road network radiating from the centre and a circular ring road. This town has several road junctions which will be to some extent centres of accessibility, but not so important as the town centre. The highest land values are again in the town centre, but subsidiary peaks of land values exist at the junctions of the ring road with each of the radial roads. We might therefore expect the town to have a central core of shops (high land values) surrounded by an industrial zone, but these would probably extend outwards along the radial roads. At each of the road junctions along the ring road we might expect there to be a cluster of shops and/or industry. The remainder of the town (low land values) might be mostly residential.

In buildings of two or more stories, there is a clear difference in accessibility between the ground floor and the upper floors. Even in our homes this difference in accessibility is reflected in the way we use the house. When sleeping for example, we have no need for accessibility. Hence we sleep on the upper floor which is less accessible.

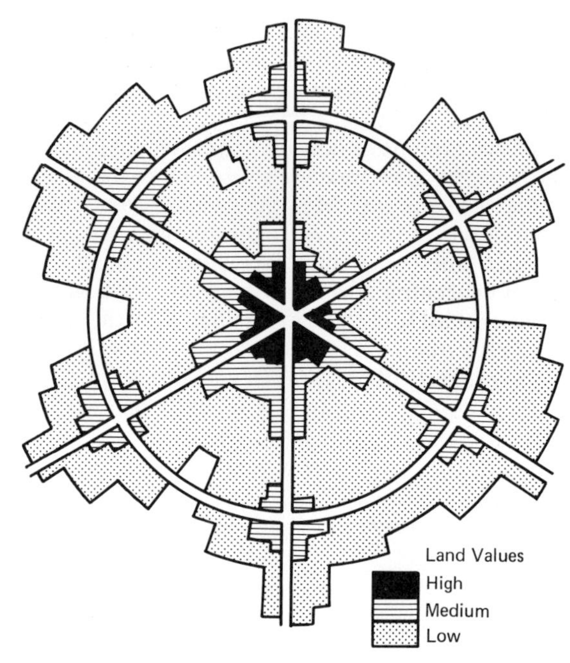

Fig. 8.13 The influence of transport routes upon land values

In commercial and industrial buildings the same principle applies. An industrial building may have a workshop on the ground floor and a store room above. In town centres offices often occupy second or third floor rooms, above ground floor shops. The same sequence as is explained above from the town centre to the outskirts may occur from the ground floor to the top floor of a building: shops on the lower floors, industry (storage) above the shops and residence (penthouse flats) on the highest floors.

The invention of the lift has greatly increased accessibility to the upper floors of buildings, in a similar way to a radial road from the town centre to the outskirts. This has allowed the shopping function to extend upwards in buildings. With staircases only, shops are probably limited to two effective floors, as with small shoe shops with men's and women's departments on different floors. With lifts, it has been possible to create multi-storey department stores.

Various models of urban land use patterns have been proposed by geographers. These are generally based on three principles: that land uses are arranged *concentrically* around the town centre (as in Fig. 8.12); that they tend to arrange themselves in *radial sectors* each running from the centre to the outskirts; and that land uses are arranged around certain centres of high accessibility within the town which act as *nuclei.*

Burgess presented his concentric model in the mid-1920s. It was based upon studies of Chicago at that time. He envisaged that various land uses form concentric zones around a central business district (CBD) in the town centre. Closest to the CBD is a *transition zone* containing industries and slum property often occupied by immigrants. The next ring outwards contains the homes of *industrial workers*, conveniently near the industries of the transition zone. Further out still is a zone of *better class residences*, and this is succeeded by the zone where *commuters* to the town live, up to one hour's travel time from the central business district.

Hoyt in the late 1930s offered an alternative suggestion, based upon studies of a large number of American cities. He concluded that land uses tend to be arranged in *sectors* radiating from the town centre. As the town grew, a better class residential area, for example, would tend to expand outwards and thus create a radial sector of better class residences all the way from the town centre to the outskirts. Similarly, industry would tend to be found all along a radial railway line or a canal or a river.

In the 1940s Harris and Ullman proposed a different model. In this case land uses would tend to cluster around a number of different centres of high accessibility in addition to the central business district, thus giving a much more complicated land use pattern. There would be outlying shopping centres in addition to the central business district. Some industries would cluster near the docks or the railway yards. Similar and related land uses would cluster together, for example low class residence and industry, various kinds of offices. Dissimilar and incompatible land uses would remain far apart, such as industry and high class residence. Thus a patchwork of land uses would be produced with no clear concentric or radial tendencies.

These models are generally inadequate to describe or explain the characteristics of modern towns, but they are useful as standards with which to compare present-day land use patterns. Also it is profitable to investigate the importance of the various factors mentioned in these models in shaping the very complex land use patterns of towns.

Urban land use patterns in the cities of developing countries often differ greatly from those in advanced countries. Movement within and between cities is less important. Hence the importance of accessibility is less than in cities in advanced countries. In some cases, the central business district is only poorly developed. The town centre is often largely occupied by churches or administrative buildings. Also, the better class houses, occupied by rich and powerful families are often located in the centre of the town and social status tends to decline towards the outskirts where, frequently, low class shantytowns exist.

Very often it is difficult to identify clear examples of shopping or industrial zones within the town. Much industry is at the small-scale, workshop level and may be conducted on the ground floor of ordinary residential houses, the 'industrialist' living above or behind his workshop. Similarly many shops exist in the form of 'shop houses' as the ground floor front room of residential property. Such workshops and shops may be scattered haphazardly through residential areas. Any large scale industry tends to be located outside the town, near to either a railway station or, increasingly, the airport.

Because it is difficult to identify clear land use zones in the towns of developing countries it usually is not possible to establish clear relationships with the models described above, which were designed in relation to the cities of the United States. As these cities become westernized they tend to acquire some features of the land use patterns of European and North American cities.

The characteristics of the different urban land uses

Towards the end of nineteenth century a central business district composed to a great extent of shops began to be created in the centres of British towns, often occupying buildings which previously had been used for residence. As the town grew, shops were also set up in the new residential areas. These are now known as suburban shopping centres. They range from small groups of four or five shops to recently developed pedestrianized shopping centres with car parks.

Before the widespread use of the motor car, public transport services in British towns (tramcars and buses) ran mainly along the system of radial roads, linking the residential areas with the town centre. Hence 'ribbons' of shops grew up along these roads, especially near the fringes of the central business district.

Shops may be roughly divided into two main types. There are those that sell *shopping goods* (goods that are bought infrequently and that cost a considerable amount of money, e.g. items of furniture) and *convenience goods* (goods that cost relatively little and that are bought frequently, e.g. most food products). Shops that sell shopping goods have a higher *threshold* than those that sell convenience goods. They need to have a larger catchment area of possible buyers. Most convenience goods shops can survive with fewer buyers shopping more frequently. Hence shopping goods shops tend to be concentrated in the town centre where they can draw buyers from the whole town. Convenience goods shops are scattered through the rest of the town nearer to their customers but with smaller catchment areas. However, shopping goods shops are found in large suburban shopping centres.

Generally in British towns, suburban shopping centres are spaced in relation to population density. They tend to be closer together in the densely populated areas near the town centre and to be further apart in the sparsely populated outer suburban areas.

Shop distributions also tend to reflect the variations in wealth between households in different parts of the town. Second hand furniture shops and cheap clothing shops are usually more common in the inner parts of a town, and antique shops, boutiques and restaurants in the more affluent outer areas.

The increasing use of the motor car instead of public transport has caused changes in the shop patterns of British towns. Many of the ribbons of shops along radial roads near the town centre have decayed, sometimes partly through the introduction of parking restrictions to speed the traffic flow.

In North America, where car ownership levels are very high, even the central business districts have decayed. Because of traffic congestion and parking restrictions, the CBD is no longer the most accessible location for a car-owning population. Hence even large department stores have moved to huge shopping centres on the outskirts.

In Britain attempts have been made in many towns to make the town centres attractive and accessible by the provision of traffic-free pedestrian precincts and multi-storey car parks often located near large enclosed, heated shopping centres.

In the last 20 years in Britain supermarkets have greatly increased in number. Because more households have adequate storage facilities for food it has become more common to shop less frequently for convenience goods. Also, the increase in car ownership has made bulk shopping much easier than it ever could be if public transport were used. Early supermarkets were often set up in town centres, but the greatest developments have since taken place near the outskirts where more land is available for car parks and there is less traffic congestion. Large supermarkets benefit from *economies of scale* and lower prices are charged. The turnover per square metre of selling space is much greater than in a traditional shop and economies in the use of labour are made by using the self-service system whereby some of the work involved in running the shop is transferred to the customer. Supermarkets on a town's outskirts tend to draw many of their customers from the town's outer suburban areas where families are generally fairly affluent and car ownership is at a high level. Thus a supermarket's catchment area tends to extend round the circumference of the town and far into rural areas outside the town, particularly along ring roads and other roads along which quick journeys can be made.

Offices in general are concentrated in the central business districts of towns to a greater extent than any other type of land use. They use space intensively and can be located in relatively inaccessible positions off the main shopping streets and sometimes in the upper stories of buildings.

Functional linkages are extremely important in office location. There are many advantages to be gained by clustering in the central business district. Many offices need the services of other offices such as those of solicitors and accountants, and they also need the services of printers, who are commonly located in town centres. Some offices, particularly banks and those of estate agents need to be located in main shopping streets in the town centre so as to be convenient for customers. Also in the town centre it is common to find clusters of solicitors near the law courts and of stockbrokers near to the Stock Exchange.

The head offices of very large firms are commonly located in the centre of large cities, particularly in well-known streets so that their address may possess a considerable amount of prestige. These offices are used by the highest level of management who find it important to meet other senior managers face to face rather than communicating by telephone or letter. These meetings often take place informally over lunch at some city centre restaurant. These contacts are especially important in the fields of finance, entertainment (e.g. television), the clothing industry (fashions) and publishing.

Other offices are found frequently outside the central business district. Those which provide a personal service to customers (e.g. doctors, dentists) are distributed in a similar way to the distribution of population densities in the town, each one having a relatively small catchment area immediately surrounding it.

Sometimes offices which are chiefly of the 'town centre' type, such as estate agents, travel agents and solicitors, are located in or near shopping centres in the high-income residential areas of the town.

At the present time there is a tendency for some offices to move out of town centre locations to cheaper land on the outskirts of the town. Routine office work is becoming more and more automated. Hence, there is less need for a large labour force to commute daily, but office functions which require face to face contact tend to remain in town centres. In some cases offices may locate near an airport. Air transport is rapidly replacing rail transport for senior managers.

Industries used to be more common in town centres than they are now, but gradually they have been replaced by shops. On town centre fringes it is quite common to see an old industrial building now being used as a shop, frequently a discount store.

Some industries still remain in or near the town centre (e.g. in the transition zone (page 165)). Three common ones are printing, the clothing industry and baking. The printing industry includes newspaper publishing and commercial printing. Both may be found in less accessible parts of the town centre, frequently in streets slightly away from the main shopping streets. Newspaper publishing deals in a commodity (news) which is highly perishable both as a raw material (when it is being collected by reporters) and as a product (when it has been printed in the newspaper). Hence, accessibility to both supplies of news and markets for the newspaper are important. Much news originates in the town centre (through meetings of various kinds and entertainments) and the town centre is regarded as a convenient place both to collect news from the surrounding area and to distribute newspapers to newsagents' shops. Commercial printing gains an advantage from a town centre location through nearness to customers with whom face to face contact can be important.

Wholesaling (warehousing) also gains a considerable advantage from a town centre location. Wholesaling forms the link between manufacturing and the retail trade. It would usually be inefficient for shopkeepers to place their orders directly with manufactures. This would involve many small consignments of goods being sent to individual shopkeepers over the whole country, and each shopkeeper might have to deal with a large number of different manufacturers. Hence, wholesalers receive bulk consignments from manufacturers and distribute small lots to retailers. It is therefore an advantage for wholesalers to be located near the town's largest cluster of shops in the central business district.

Bakeries also gain an advantage from a central location in the town. As in newspaper publishing, the products of a bakery are perishable and need to be supplied quickly to the markets. From a central location a bakery can supply the whole urban market relatively efficiently.

In many cities industry in the central districts has tended to decline. Near the town centre, in the transition zone, buildings are often old-fashioned and unsuitable for modernisation. It is sometimes difficult to build a completely new factory because of the fine-grained street network which is adapted only to small buildings. Also, the increase in road traffic has led to congestion in these central areas, which reduces the advantages of accessibility that might be expected to result from a central location within the town. Also, as people have moved out of the inner districts of towns, industries have lost the advantage of a nearby labour supply.

In many British towns large-scale clearance projects have been carried out. Old decaying property has been demolished and the ancient street plan has been replaced by a new system of roads. Wholesalers, car sales firms (which tend to use space extensively) and modern manufacturing industries have set up on these new industrial estates in the transition zone, close to the central business district.

Modern space-extensive industries are more often located near the outskirts of the town where open space is more plentiful and land prices are lower. Also, road transport facilities for large lorries are better than in the town centre. Particularly valued locations are near a motorway intersection which gives easy access to regional and national markets. Large industrial concentrations have developed at a number of major junctions in the British motorway network in what until a few years ago was open countryside. These industries are also attracted to out-of-town locations with railway transport facilities.

Industries that use heavy or bulky raw materials are often located where efficient bulk transport facilities exist, as near docks or railway junctions.

In the past hundred years or so the *pattern of population densities* in the cities of advanced countries has changed greatly. As these cities grew through the nineteenth century the population density increased at the centre as the city increased in size (lines 1, 2 and 3 in Fig. 8.14(*a*)). As the central business district developed, the population density decreased at the city centre (lines 4 and 5 in Fig. 8.14(*a*)). Meanwhile the city was expanding as outer suburbs were created (lines 4 and 5) and, as the motor car came into common use the population density gradient towards the city's fringe became very gentle. Urban sprawl had begun to cover the countryside.

These different population densities within towns are reflected in the different house types that exist. Near the point of highest population density, not far from the town centre, the houses are usually small, terraced, with very little, if any, garden space. On the outskirts it is normal to find that the garden area is at least as large as the area of the house. Houses are more widely spaced.

In developing countries, urban population density patterns are usually quite different. As population densities have increased with urban growth they have continued to grow in the town centre, and there has usually been little tendency for a central business district to appear. Also, the gradient of the decline in population density towards the outskirts has tended to remain constant (Fig. 8.14(*b*)). Urban sprawl usually does not occur, but some cities, such as Bombay and Calcutta, founded by the British, have tended to develop central business districts in the same way as those in advanced countries. Hence an area of low population density exists at the city centre.

Socio-economic aspects of population also vary in different parts of towns. In advanced countries it is usual to find richer people and more expensive houses in the outer suburban areas than in the inner areas of towns. People with low incomes tend to live nearer to their work. It is common to

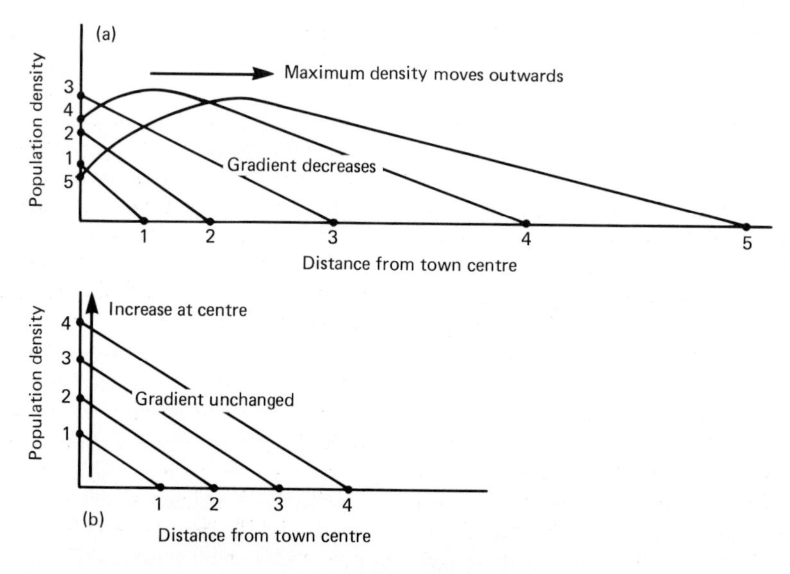

Fig. 8.14 The evolution of urban population density patterns

find low-income housing very close to the town's industrial areas. People with high incomes have a greater choice of location for their homes. They can choose areas with attractive scenery at the outskirts. Since they have a high level of car ownership they are highly mobile and need not consider the costs of the journey to work. In some cases extremely wealthy people may own a town house in the town centre for use during working time and a country house well outside the urban area for weekends.

Moving from the town centre to the outer suburbs, it is usually found that the average age of the population decreases and the number of children in the family increases.

As time passes there is a tendency for houses to 'filter down' to families of lower income. As the age of the house increases, wealthy families tend to move to newer houses further from the town centre, and they are replaced by other families moving from inner parts of the town.

In the large cities of the USA serious financial problems have arisen through industry and relatively wealthy people moving into rural areas on the city's fringes. The central parts of these cities are dominated by relatively poor people who are unable to pay enough in taxes to maintain essential services, such as health and fire protection in the central city. In Britain the reorganization of local government boundaries in 1974/5 often combined the poorer inner cities with the more affluent outer suburban areas.

In the cities of the USA, and to some extent the cities of England, immigrants have tended to occupy the poorer quality housing in the inner areas. Thus, ghettos of considerable size have developed in many large American cities (page 23).

In many cities in developing countries the situation is completely different. Here, the richer people tend to live near the city centre and the poor on the periphery, sometimes in shantytowns. As a middle class develops however, there is a tendency for suburbs in the Western style to grow up.

London has many serious inner city problems. The population of the 'City' where so many offices are now clustered is only about one-tenth of its total in the middle of the nineteenth century. Industries have also moved out of the inner parts of London. This has left a population of generally poor people, including many immigrants, who have an unusually high rate of unemployment.

The central business district

The central business district is the greatest concentration of shops and offices in a town. It also has the highest land values and usually the tallest buildings, especially in American cities.

In British towns the central business district developed often in the late nineteenth century near the major road junction at the town centre, on land which had previously been occupied by residence or small-scale industry. It usually occupies the oldest part of the town, but its buildings are constantly being rebuilt or altered. Commonly the ancient, fine-grained street plan survives but sometimes it has been destroyed as large shopping centres and office blocks have been built.

The size of the central business district varies with the town's population. In a small town it may not exist as a compact unit. Instead the rows of shops and offices may be interrupted by residences and even small industrial buildings. In larger towns the shopping area becomes more compact and a separate office zone may develop. In very large cities this separation of functions is more complete. Central London, for example, has two major shopping areas, the West End and Kensington, separated by a belt of parks. Certain parts of these shopping areas also specialize in particular commodities such as women's clothing, jewellery etc. London also has two major office zones, the City (banking, insurance and trading) and the Whitehall district (political offices), at opposite ends of the Strand/Fleet Street.

The shopping area of the central business district has the highest land values and the greatest densities of pedestrians moving along the pavements. Offices tend to be tucked away in less accessible positions behind the shops where land values are lower. Offices do not depend upon a high level of visibility to pedestrians. Alternatively, offices cluster in high-rise blocks often built behind the main shopping streets. Shops frequently cluster around the point of highest land value in the town (the peak land value intersection) but they may extend as ribbon development towards the railway station, the bus station or an important, permanent car park.

Large shops such as department and variety stores often cluster near the peak land value intersection. These depend upon high density pedestrian flows and tend to generate such flows. Clustering makes it convenient for shoppers to visit the whole group, to the benefit of each shop.

The central business district of Bolton

Women's clothing and shoe shops are usually located near these large stores to take advantage of the heavy pedestrian flows. Many of these shops may compete for corner sites which give the advantage of extra visibility to pedestrians. In this central part of the CBD shopping space is used very intensively with a high value of sales per unit of shop space.

Further away from the peak land value intersection furniture shops appear. These need a larger amount of display space and space is used less intensively. Towards the margins of the CBD, food shops such as grocers' begin to appear and, at the margin, discount stores may have been established where more space is available.

Thus, there tends to be a progression from clothing and shoe shops on high value land to furniture shops on land of lower value. This progression is also common as one moves from lower, more accessible, floors in a department store to higher, less accessible floors. Women's clothing is often sold on the ground floor, and furniture is relegated to the highest floor of all.

Small specialist shops, such as jewellers, antique shops and some bookshops are often located in small side streets, off the main shopping areas. They have less need to be visible to as many pedestrians as possible because their trade does not depend so much upon impulse buying. People make special trips to visit these shops.

Certain types of shops tend to cluster close to one another especially in cases where shoppers may wish to compare fashions and styles, as with shoe shops and furniture shops. Also women's shoe shops tend to locate near women's clothing shops.

Some offices also tend to locate on the main shopping streets. Some, such as banks and estate agents, are here because they have the same need as shops to be accessible to the maximum possible number of pedestrians. Other offices are commonly found on the higher floors of buildings whose ground floors are occupied by shops.

The central business district of Vancouver

URBAN SPHERES OF INFLUENCE

Relations between towns and their surrounding areas

Towns are linked in many different ways to the country areas that surround them. They provide many different services for the people who live in their spheres of influence. Such services include entertainments of various kinds, also educational facilities (technical colleges, libraries and schools), health facilities (doctors and hospitals) and various cultural activities (lectures and meetings of various societies). In addition the industries, shops and offices provide work for the people who live in surrounding areas and many people travel to work in the town every day. Towns also frequently act as collecting centres for the produce of surrounding rural areas. Cattle markets and other types of markets are usually found in towns.

The size and shape of a town's sphere of influence (market area or urban field) can be studied in many different ways. A study of the advertisements and the news items in a town's newspaper will sometimes give a rough idea of the area influenced by the town. Also it may be possible to discover the location of all the newsagents' shops to which the newspaper is delivered. The census of population gives details of the home district of the people who work in the town. Thus, one can define the catchment area of the labour force that works in the town. Sometimes it is possible to discover the locations of the homes of children who attend certain schools or people who attend a cattle market or who are members of various urban societies or lending libraries. Sometimes urban spheres of influence have been estimated by field work, by interviewing shoppers and discovering the general location of their homes. In other cases, field work has been carried out in rural areas and interviews have been conducted to discover which town people usually visit when shopping for various goods and services. It is also possible to build up a general idea of a town's sphere of influence by using bus timetables. The procedure is to count the number of buses beginning at a town which pass along the various roads to the various surrounding villages in a typical day.

Usually it is found that the strength of the influence of a town over its surrounding area steadily decreases with increasing distance from the town.

The use of the gravity model to estimate the extent of the sphere of influence

According to the gravity model (page 145) the interaction between two towns decreases in proportion to the square of their distance apart. Similarly the influence of a town over its surrounding area is thought to decrease in proportion to the square of the distance from the town.

Figure 8.15(*a*) shows four towns of equal size arranged equal distances apart in a straight line. The upper diagram shows the strength of their influence over their surrounding areas. This influence in each case decreases quickly near the town, then more slowly further away from it. In these diagrams each town is the same size. In the lower diagram (map) it can be seen that each town's sphere of influence has the same area.

In Figure 8.15(*b*) it is assumed that towns A and C have increased greatly in size. The upper

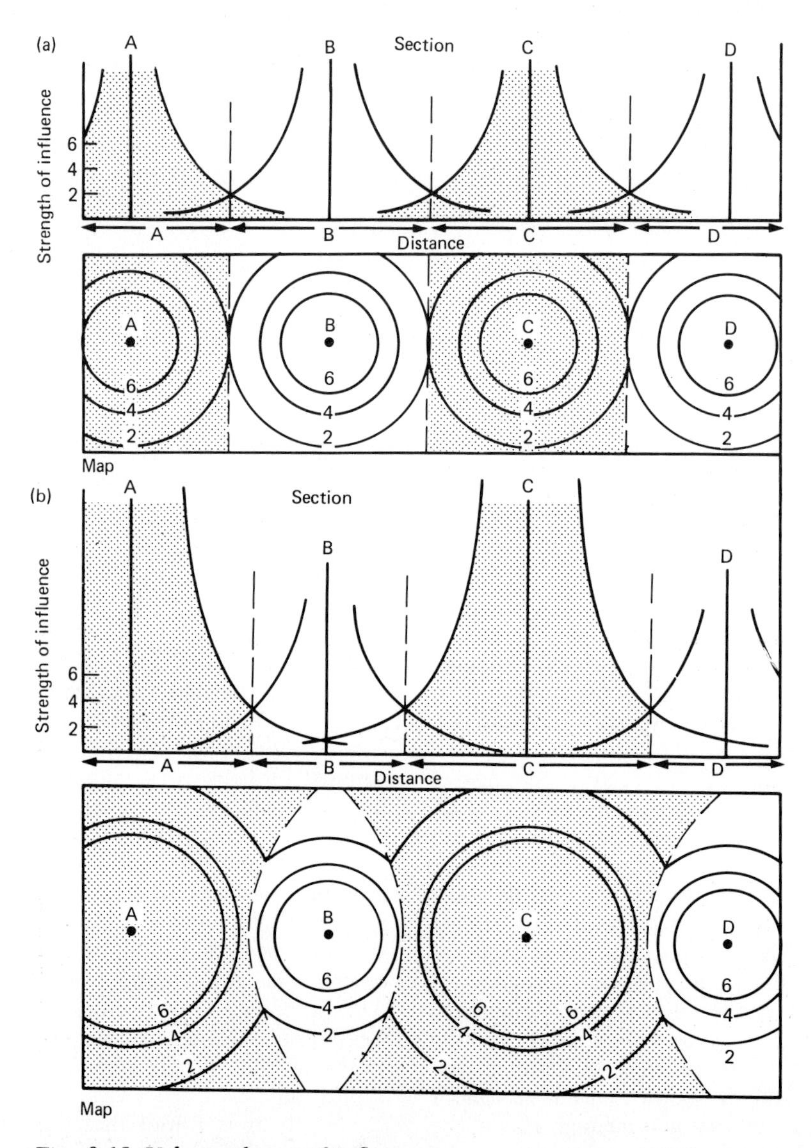

Fig. 8.15 Urban spheres of influence

diagram shows that their influence over their surrounding areas has increased greatly and that they have 'captured' part of the spheres of influence of towns B and D. This is made clearer in the lower diagram (map). Here it is clear that the spheres of influence of towns A and C have made B's sphere of influence much narrower than before. The larger spheres of influence have also tended to 'wrap themselves round' that of towns B and D.

Figure 8.16(*a*) shows seven towns (A to G) distributed evenly over an area. All these towns have exactly the same population total. By the gravity model therefore we would expect every town to exert an equal influence over its surrounding area. Hence, the break points dividing the towns' spheres of influence from one another will be halfway between neighbouring towns. This produces a regular hexagonal pattern of spheres of influence, and all are equal in size. Such a pattern will rarely occur in reality because it is extremely unlikely that the towns in any area will be equal in size and equally spaced.

In Figure 8.16(*b*), the seven towns are still the same size, but they are no longer equally spaced. This clearly has a great effect upon the size and shape of their spheres of influence. The sphere of influence of D for example is no longer symmetrical and D is no longer at its centre. This is because A and C are now much further away from D than are B and E. Thus, a town's location in relation to other towns has a great influence upon the shape and size of its sphere of influence. If all the other towns were located nearer to D, D's sphere of influence would be very small.

In Figure 8.16(*c*), the seven towns are again equally spaced, but this time their population totals are very different. Again D is not central to its sphere of influence. This extends much further to the north-west than to the south. This is because towns A and B are very small compared with D, so D captures some of their territory. On the other hand, to the south, G is much larger than D, so it captures some of D's territory.

It is easy to calculate the position of the break point between two towns by using the formula below. In this formula, B is a smaller town than A.

$$\text{distance of break point from B} = \frac{\text{distance from A to B}}{1+\sqrt{\left(\dfrac{\text{Population of A}}{\text{Population of B}}\right)}}$$

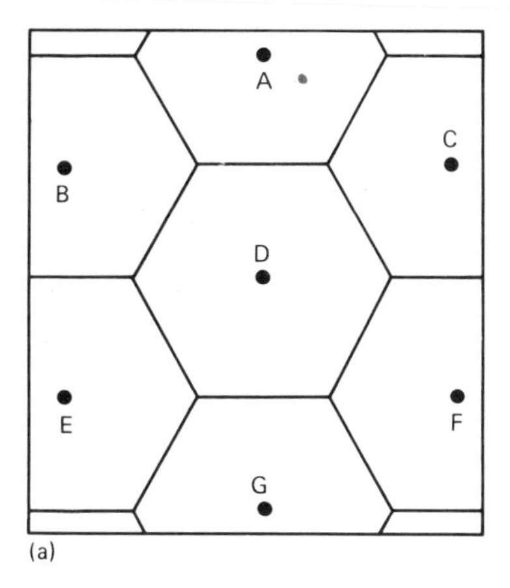

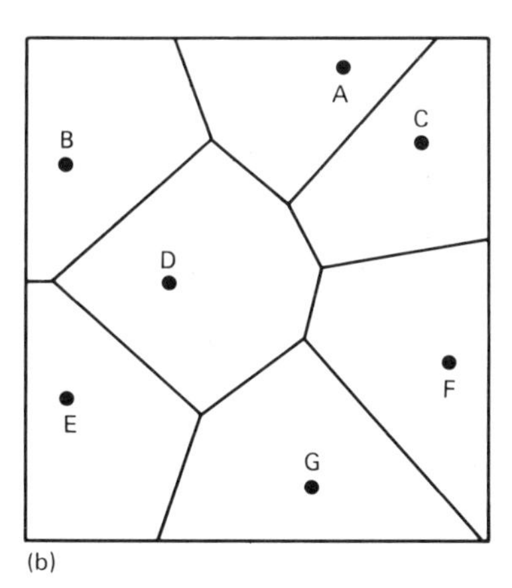

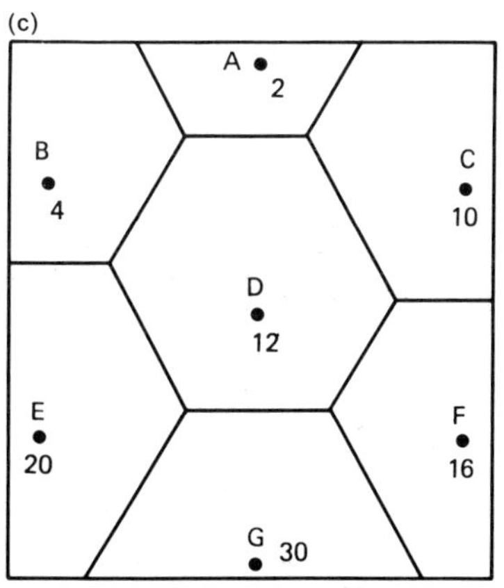

Numbers indicate population of town in tens of thousands

Fig. 8.16 The influence of spacing and population total upon urban spheres of influence

To work out an example using this formula do the following steps in turn:

(*a*) Measure the distance from A to B.
(*b*) Divide the population of A by the population of B; then find the square root of this figure. Then add one to the square root.
(*c*) Finally divide your answer in (*a*) by your answer in (*b*). This gives the distance of the break point from the smaller town measured in the direction of the larger town.

It is possible to use the number of shops in the town instead of its total population if you wish.

Although the method described above is an attempt to draw a line separating the spheres of influence of neighbouring towns, it should be remembered that a town's sphere of influence gradually weakens further away from the town. It does in fact 'decay with distance', as explained on page 172.

The journey to work

As residential suburbs have expanded on the outskirts of cities, while the greatest concentration of jobs still exists in the city centre, the journey to work has lengthened considerably. In most British towns the large-scale movement of people into the town centre in the morning and back to the suburbs in the evening has created severe traffic problems.

This radial movement is still important, but now industries, shops and even offices are moving out to the suburbs. The result is that journey to work patterns are becoming more complex, with important cross movements as well as radial movements. The journey to work patterns in Greater London illustrate this complexity.

Figure 8.17(*a*) shows that in central London (Westminster, the City of London, Tower Hamlets and Newham) a majority of the working population works within the local borough. This is also

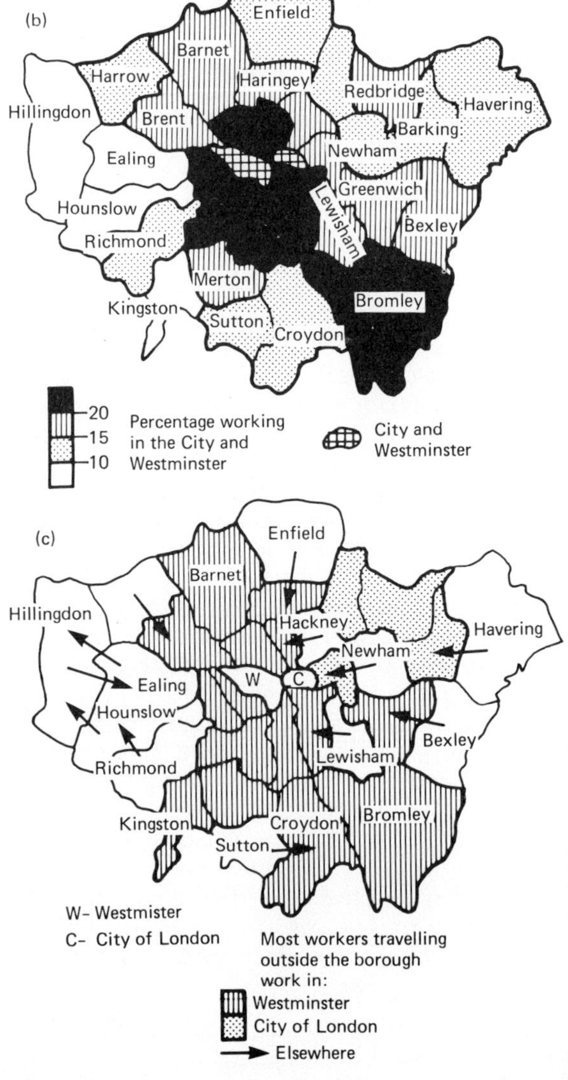

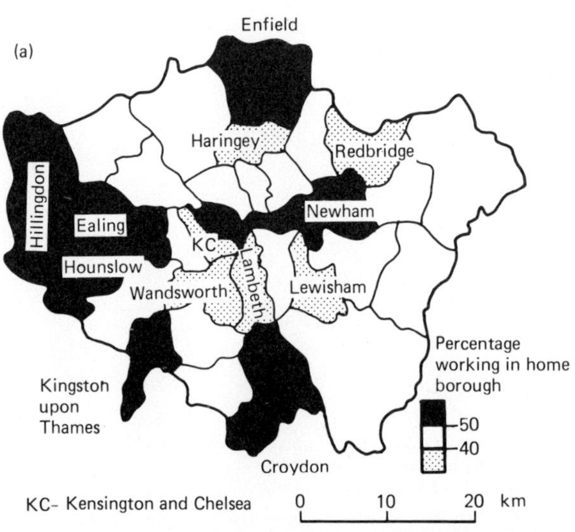

Fig. 8.17 The journey to work in Greater London

the case in Enfield to the north, Croydon and Kingston-upon-Thames to the south, and Hillingdon, Hounslow and Ealing to the west. In contrast, relatively few of the residents of an inner ring of boroughs find work locally. This ring consists of Redbridge and Haringey to the north and Kensington and Chelsea, Wandsworth, Lambeth and Lewisham to the south.

Figure 8.17(*b*) shows the extent to which central London (the City and Westminster) attracts workers from the rest of the conurbation. Its strongest influence is concentrated in a north-west to south-east belt extending to Bromley. The far north-east (Havering) and the west and south-west (Hillingdon, Hounslow, Ealing and Kingston-upon-Thames) are less affected. Most workers in the latter area are employed locally (Fig. 8.17(*a*)).

Figure 8.17(*c*) shows the destination of the journey to work of the greatest number of people who travel to work outside their home borough. Westminster with its concentrations of shops and offices is a strong attraction in almost all the boroughs from Barnet in the north-west to Bromley in the south-east. The City of London has a different catchment area, mainly to the north-east in Tower Hamlets (nearby) and Waltham Forest, Redbridge and Barking (further out). There are also considerable movements between neighbouring boroughs, as between Ealing, Hounslow and Hillingdon. Work opportunities in Hillingdon, in particular, seem sufficient to rival the attractions of Westminster for residents of Ealing and Hounslow. This movement is *outwards* from the centre of the conurbation.

Unlike Greater London, some of the other conurbations of Britain, such as Greater Manchester and West Yorkshire, are made up of fairly free-standing towns surrounding the central city. The city and each of the free-standing towns are surrounded by suburban residential areas. Figure 8.18 shows an idealized pattern. Two main kinds of work journeys are made by suburban residents. Some travel radially inward to the centre of their local town; others make the longer journey to the central city. These journeys are made feasible by the high level of car ownership among the suburban families. In Greater Manchester an additional feature is the radial belt of industry along

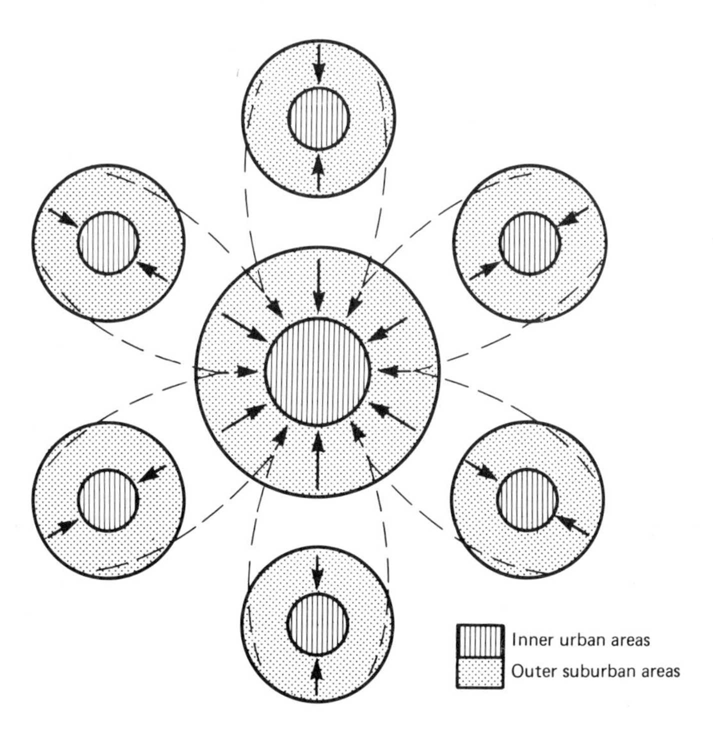

Fig. 8.18 Theoretical journey to work pattern in a conurbation

the Manchester Ship Canal (including Trafford Park) to the west of the city. This attracts workers from the towns of the outer ring, and journeys are helped by the motorway (M62/M63) around the western outskirts of Manchester.

Central place theory

Central place theory was introduced in a book written in the 1930s by Walter Christaller, based upon studies of settlements in Germany. It is concerned with the patterns of cities, towns, villages and hamlets in an area and the ways in which they provide various services (such as wholesale and retail trade, legal, medical and banking facilities, entertainment and administration, for example) to the people who live in their surrounding areas (spheres of influence or market areas). It is concerned with the locations of such service centres and the shapes and sizes of their spheres of influence. A settlement which provides services for its surrounding area is termed a *central place*.

In central place theory various assumptions are made, as in the von Thünen model (page 65) for agriculture, and in the Weber model (page 115) for industry. It is assumed that the central places are located on a limitless plain over which movement is equally easy in all directions and where population and spending power are uniformly distributed. All the inhabitants of the plain keep their travel distance to a minimum by obtaining any service at their nearest central place which provides it. All locations on the plain are able to be served by central places. Transport costs over this isotropic plain vary in direct proportion to travel distance. Since population and spending power are evenly distributed, central places are evenly spaced over the plain.

The *range* of a service is the maximum distance that an inhabitant of the plain will travel to obtain a service at a central place. This is explained in Figure 8.19. The left-hand graph shows that the price of a service *at* the central place itself is OP. The dotted line running from P to the right-hand graph shows that, at this price, if the buyer lived at the central place, he would buy quantity OA. However, further away from the central place, transport costs have to be paid (shown by line PY in the left-hand graph). At distance OX from the central place, the buyer would have to pay XY for the service. XW (or OP) of this would be the actual price and WY would be transport costs. The right-hand graph shows that at price XY (which equals OQ) the buyer would not be willing to buy the service. Hence, distance OX is the *range* of this service. This can be regarded as the radius of a circle centred on the central place.

High order services have a larger range than low order services. For example, a university draws its students from a much wider area than a village primary school.

The *threshold* of a service is the minimum amount of sales or the minimum number of buyers needed to make it worthwhile to provide the service. It is clear that it would take a much larger population to justify setting up a university than would be needed to set up a small primary school. Hence, on our isotropic plain, universities would have to be located further apart than primary schools, and there would be fewer of them. If there were too many universities some would have to close down because there would not be enough

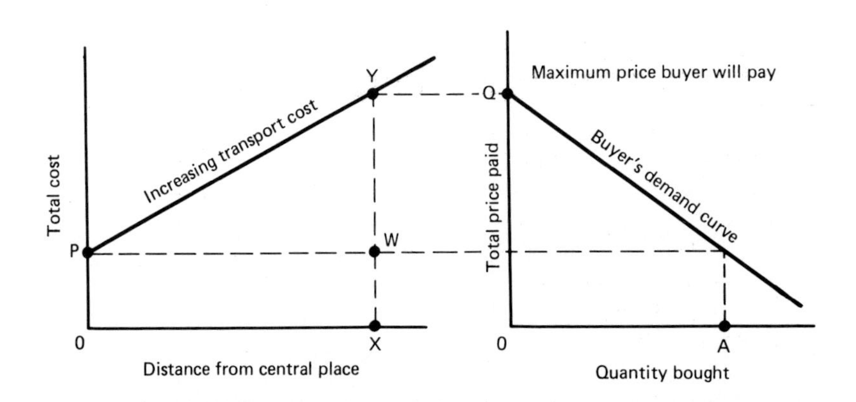

Fig. 8.19 The range of a service

students to reach their threshold. High order services like universities have a higher threshold than low order services like primary schools. Fortunately for them they also have a larger range (see above). Otherwise they could not exist.

In central place theory, central places are distributed over the area in such a way as to serve the evenly distributed population as efficiently as possible. The ideal shape of a central place's sphere of influence is circular. This is because a central place is nearer to all parts of a circular sphere of influence than it would be for any other shape of the same area. If these circular spheres of influence are packed as closely together as possible (Fig. 8.20 (*a*)), this results in a *triangular* pattern of central places. However, taken as a whole the system shown in Figure 8.20 (*a*) is not very efficient because it leaves some areas (shaded) unserved by any central place. This problem is solved by overlapping neighbouring circles (Fig. 8.20(*b*)) and dividing the areas of overlap equally between the two nearest central places. This gives a set of hexagonal spheres of influence (Fig. 8.20 (*c*)). A regular (all sides equal) hexagon is the nearest shape to a circle that can be used to cover an area completely. This is because its internal angles are all 120 degrees, so that three of them will fit together snugly (Fig. 8.20(*d*)). A hexagon is almost as efficient as a circle in terms of accessibility of its centre to its outer boundary, and it is much more efficient than a square or an equilateral triangle of the same area.

The hexagonal arrangement can also provide spheres of influence (market areas) of various sizes: large ones for *high order* (high range and threshold) and smaller ones for low order central places. This is done by interlocking the various sizes of hexagons with one another in different ways (Fig. 8.21).

Figure 8.21(*a*) shows one of these ways. In this diagram the third (highest) order central place is at the centre of a large third order market area; each of the six second order central places is at the centre of a smaller second order market area; and each of the 24 first order central places is at the centre of a small first order market area. Figure 8.21(*a*) shows that the third order central place is also at the centre of a second order and a first order market area. This is because third order central places provide second and first order services as well as third order services. For a similar reason each second order central place is at the centre of a

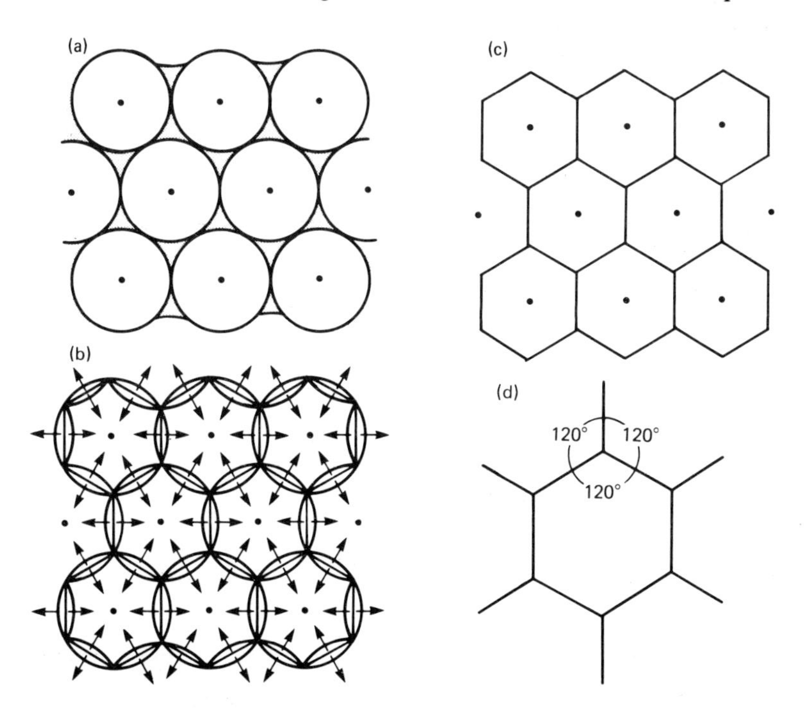

Fig. 8.20 A triangular-hexagonal pattern of central places and their spheres of influence

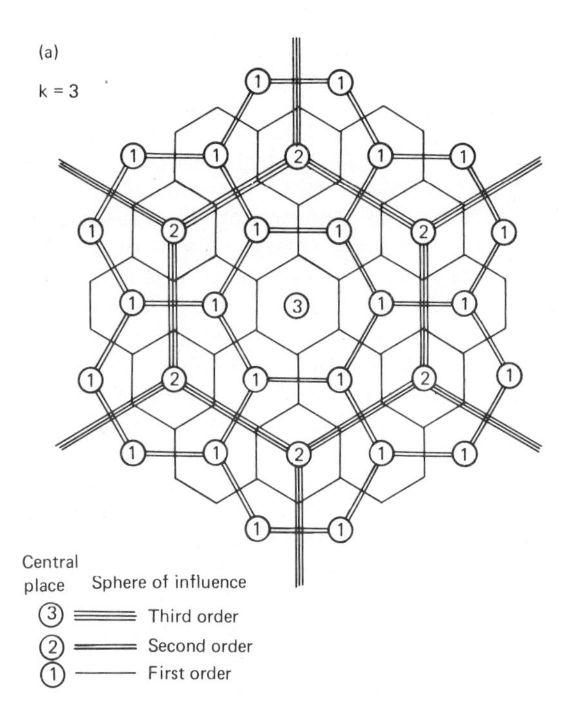

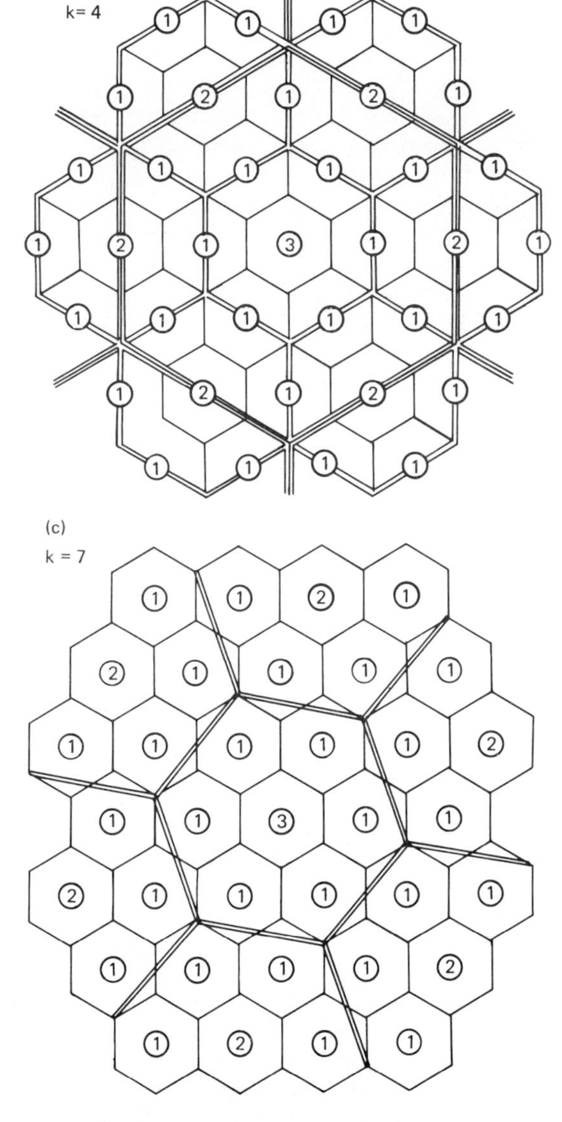

Fig. 8.21 Three different systems of central places

first order market area as well as one of the second order. This arrangement is known as the $k = 3$ system. In Figure 8.21 (*a*) it is easy to see that second order market areas are *three times* as large as first order, and that third order market areas are *three times* as large as second order.

Figure 8.21(*b*) shows a different arrangement. In this case second order market areas are *four times* as large as first order, and third order market areas are *four times* as large as second order. This is known as the $k = 4$ system.

In Figure 8.21(*c*) it is easy to see that second order market areas are *seven times* as large as first order. This is therefore known as the $k = 7$ system. It has not been possible to show the third order market area in Figure 8.21(*c*) because in this case it is so large (seven times as big as the second order).

This difference in size of market areas in these three systems is shown in Figure 8.22. Here, based on the same first order market areas, the second order market areas increase greatly in size from the $k = 3$ to $k = 4$ to $k = 7$ system.

The ratios of the numbers of central places of the different orders on a limitless plain also varies between these three systems. In Figure 8.21(*a*) the central third order market area would be repeated over and over again on a limitless plain, so if we find the ratio of the numbers of central places of different orders for this third order market area it must be the same as the ratio for the limitless plain. In this third order market area there is one third order central place. Second order central places are at the junctions of three different third order market areas, so we can only count one-third of each for the one we are considering. This makes two second order places. Finally, there are six first order central places. This gives nine central places altogether. The constant of three ($k = 3$) is clear

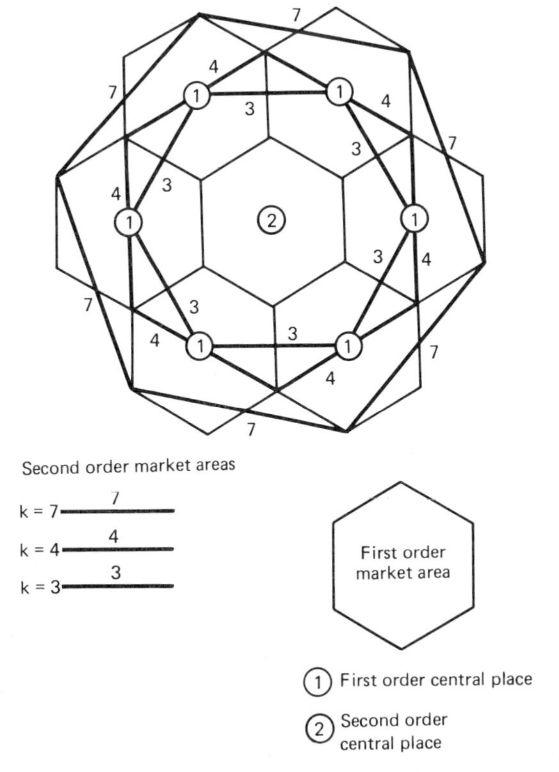

Fig. 8.22 Comparative sizes of market areas

when we reason as follows. There are nine central places altogether. Divide by three. This gives three central places of second and third order and *leaves six as first order*. Divide by three again. This gives *one central place in the third order* and *leaves two as second order*. The $k = 3$ system is said to be based on the *marketing principle*. The central places are as near together as possible, so they are as near as possible to the people who live in their market areas. In fact, in this system each successive higher order of central places is only $\sqrt{3}$ *(1.732) times further apart* than the next lower order.

In Figure 8.21(*b*) ($k = 4$), in the third order market area shown there is one third order central place, three second order central places (a half share of each of six) and 12 first order central place, making 16 altogether. This time divide by four ($k = 4$). This gives four central places of second and third order and *leaves 12 as first order*. Divide by four again. This gives *one central place in the third order* and *leaves three as second order*. Central places in the $k = 4$ system are located according to the *traffic (transport) principle*. The higher order places are located along straight lines joining the centres of the highest order, thus making it easy to link them together by means of straight roads. In this system each successive higher order of central places is $\sqrt{4}$ *(2) times further apart* than the next lower order.

In Figure 8.21(*c*) ($k = 7$) in a complete third order market area there would be one third order central place, six second order central places and 42 first order central places, making 49 altogether. This time divide by seven ($k = 7$). This gives seven central places of second and third order and *leaves 42 as first order*. Divide by seven again. This gives *one central place in the third order* and *leaves six as second order*. Central places in the $k = 7$ system are located according to the *administrative principle*. The lower order central places are never located on the boundaries of higher order spheres of influence. The higher order spheres of influence *completely enclose the lower order central places*. Thus, loyalty is not divided between two or more higher order places. In this system each successive higher order of central places is $\sqrt{7}$ *(2.646) times further apart* than the next lower order.

We should not expect to find a *perfect* example of any pattern of central places in the real world. The perfect pattern can only exist in the ideal world described by the assumptions which underlie the model (page 176).

In practice movement is not equally easy in all directions, but rather it is channelled along the links of transport networks, which may give an advantage to towns situated at transport junctions. Nor are population and spending power uniformly distributed. Population density and spending power are much greater in concentrations of industrial towns than in rural areas.

People do not necessarily keep their travel distances to a minimum by obtaining services at the nearest central place that provides them. Very frequently low order goods and service are bought in the course of another journey, such as a journey to work or a journey to buy high order goods in a city.

Ease of transport in rural areas may be related more to the frequency of bus services than to travel distance. Tradesmen from larger towns may provide house-to-house delivery services for low order goods such as bread.

The hierarchy of settlement orders in central place theory is based solely upon the provision of central place services such as the retail trade etc. In practice many settlements owe their size and importance to other activities such as manufacturing industry and tourism which involve links with areas very much greater than their local 'central place' spheres of influence. Such settlements may grow large as a result of these links with distant areas rather than their trade within their local market areas.

Nevertheless, in certain predominantly agricultural areas, with little variation in surface relief, such as Anglesey and Norfolk in Britain and the flat plains of the Middle West in the USA, a close resemblance to perfect central place patterns has been recognized. In these areas larger settlements have larger market areas and are further apart than smaller settlements. However, perfect triangular patterns of central places and perfect hexagonal market areas are never seen.

In developing countries it is common to find a modification of central place theory in the form of *periodic markets*. In such areas the level of demand for goods and services is very low, and movement (often on foot) is very slow and difficult. This means that a very large market area is needed to reach the threshold for certain services. On the other hand the range is small because of the transport difficulties. Thus it may not be possible for permanent shops to exist.

Under these conditions, periodic markets may be held, perhaps once a week in each central place. An example is shown in Figure 8.23(*a*). A market trader travels along the route shown stopping in successive days at each numbered central place and visiting each central place once every seven days. This gives the trader an effective market area extending over all seven hexagons, which may be sufficient to reach the threshold for the service. On the other hand, his customers need only to travel a short distance. Hence each of the seven central places lies within the range of all the inhabitants of its particular local market area.

If the population of the area increases, or if incomes rise, the system may change to that shown in Figure 8.23(*b*). Here each trader can survive on the basis of a much smaller market area (two or three hexagons instead of seven). Thus, he needs to visit fewer central places and markets can be held more frequently (every two or three days). Eventually, if population and/or incomes continue to increase it may become possible for each local market area (each hexagon) to have a permanent market or permanent shops.

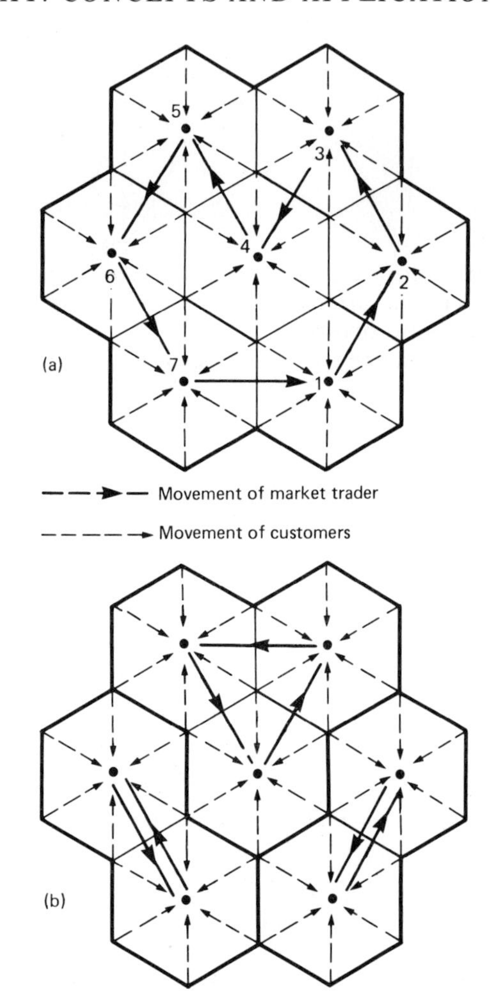

Fig. 8.23 Periodic markets in developing countries

Exercises

1. (*a*) Explain the difference between a dispersed and a nucleated rural settlement pattern.
(*b*) Quoting examples, discuss the view that in modern times rural settlements have tended to become more dispersed.
2. (*a*) Discuss the factors that can lead to urbanisation in a developing country.
(*b*) What relationships appear to exist between the process of urbanisation and (i) the demographic transition; (ii) the mobility transition?

3. The rank-size rule appears to contradict the principles that underlie central place theory. Discuss.
4. Referring to actual examples show how the morphology of cities in both advanced and developing countries constitutes a record of the history of the city's development.
5. With reference to any town or city that you have studied discuss the extent to which it conforms to the various urban land use models which have been proposed.
6. (*a*) Describe the methods that may be used to estimate the extent of a town's sphere of influence.
(*b*) Discuss the factors that may influence the size and shape of urban spheres of influence.

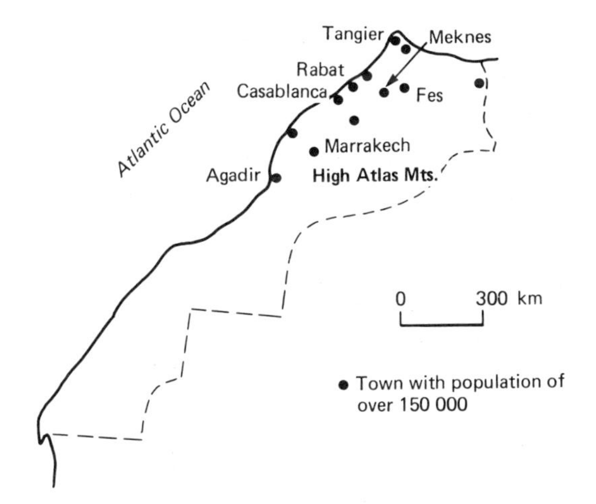

Fig. 8.24 Major towns of Morocco

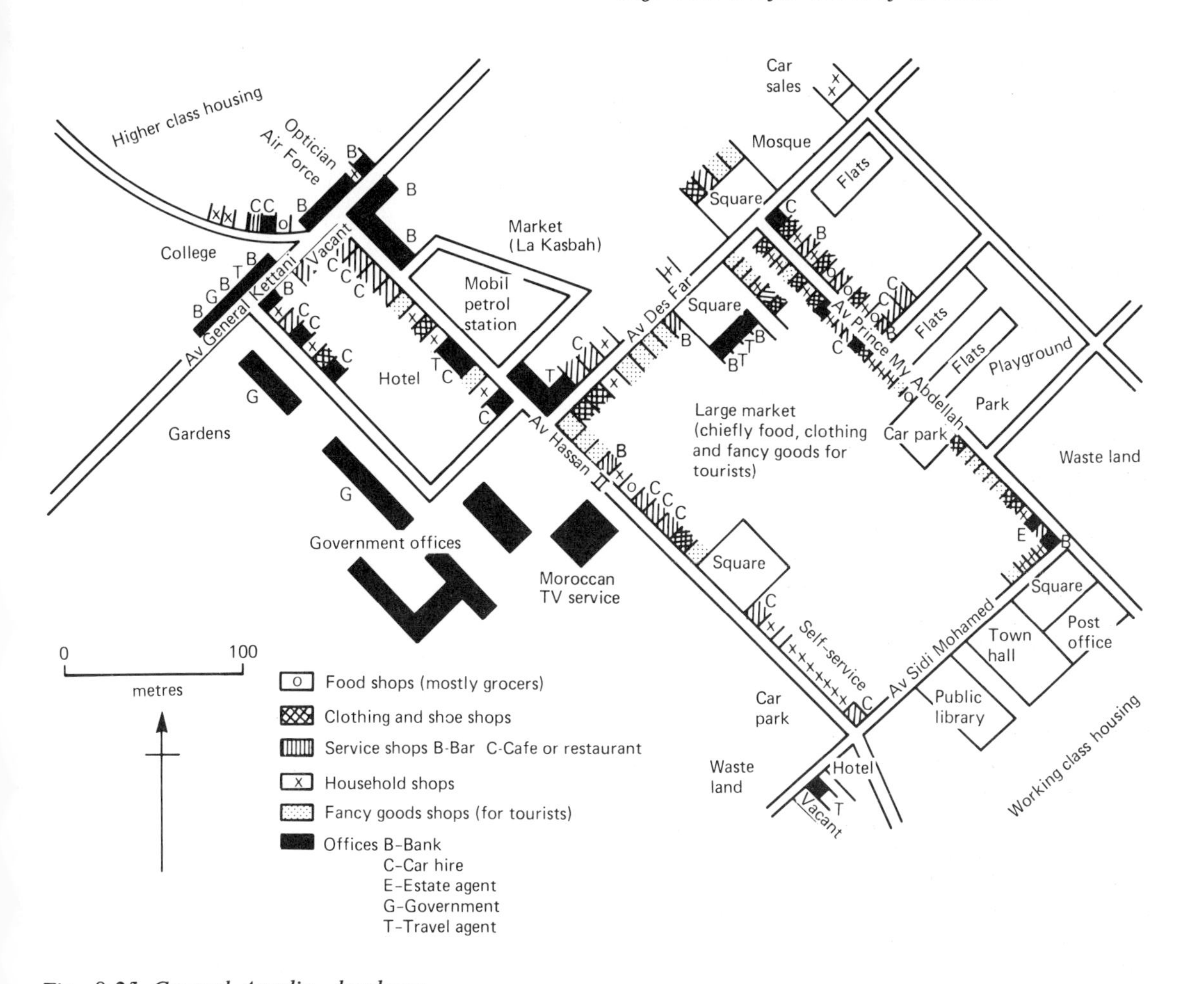

Fig. 8.25 Central Agadir—land use

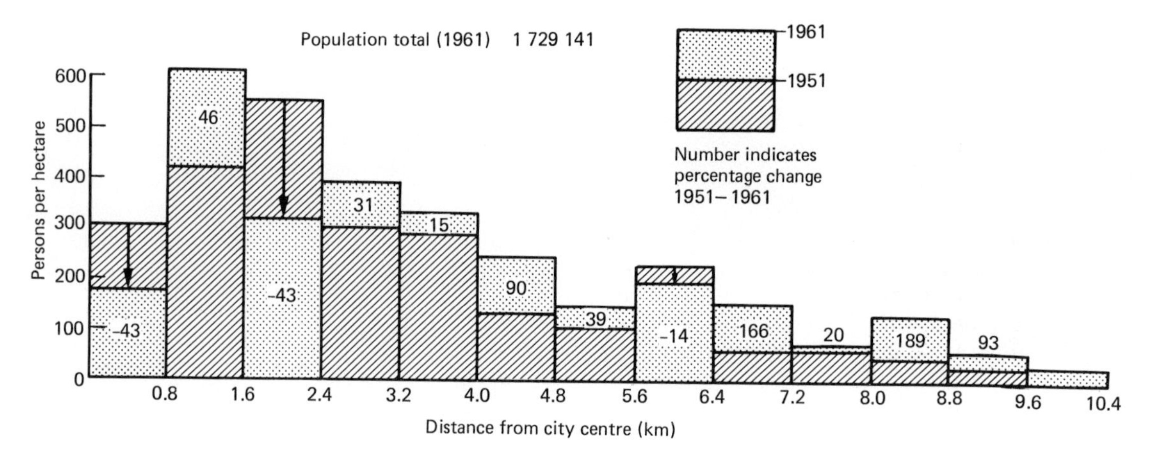

Fig. 8.26 Madras: Changes in population density 1951–1961

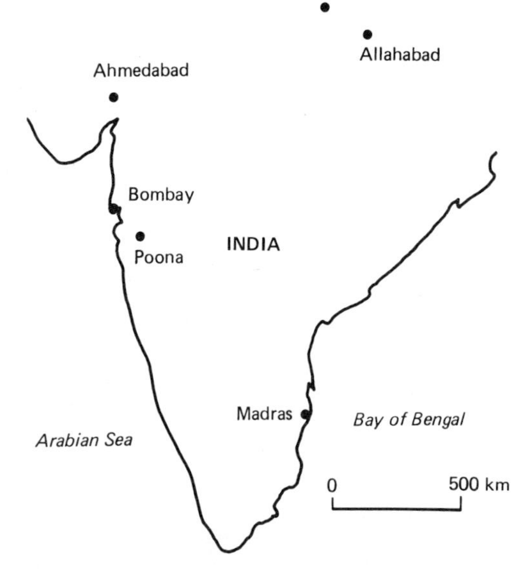

Fig. 8.27 Location map

7 (*a*) Explain the reasons why different central place systems may be governed by different constants (e.g. $k = 3$, $k = 4$, $k = 7$).

(*b*) An area contains 100 settlements of at least first order, which are distributed in a regular triangular pattern. Each settlement is located at a distance of 1 kilometre from its six nearest neighbours. For each of the central place systems $k = 3$, $k = 4$ and $k = 7$, give a description of (i) the numbers of central places in the different orders; (ii) the spacing of the central places of the different orders; (iii) the sizes of the market areas of the different orders.

8. Figure 8.24 shows the location of the largest towns in Morocco. The only large town which lies to the south of the High Atlas mountains is Agadir. Figure 8.25 shows the spatial distribution of land uses in the central part of Agadir.

(*a*) Describe and comment on the spatial distribution of the various land uses of central Agadir which are illustrated in Figure 8.25.

(*b*) The population total of Agadir is roughly equal to that of a large town in England, such as Portsmouth or Luton. In what ways and for what reasons does the central business district of Agadir differ from that of an English town of a similar size?

(*c*) In what ways do the functions of Agadir's central business district reflect its location as shown in Figure 8.24?

9. (*a*) Describe and comment on the population density gradients in 1961 of the Indian cities illustrated in Figures 8.26–8.29.

(*b*) In the light of all the data provided, discuss the extent to which the population density gradients and their development between 1941 and 1961 resemble or differ from these characteristics in European cities of a similar size.

The data in Figures 8.26–8.29 are adapted from Brush, J. E.: 'Spatial patterns of population in Indian cities' (Geog. Rev., vol. 58, July 1968) with the permission of the American Geographical Society

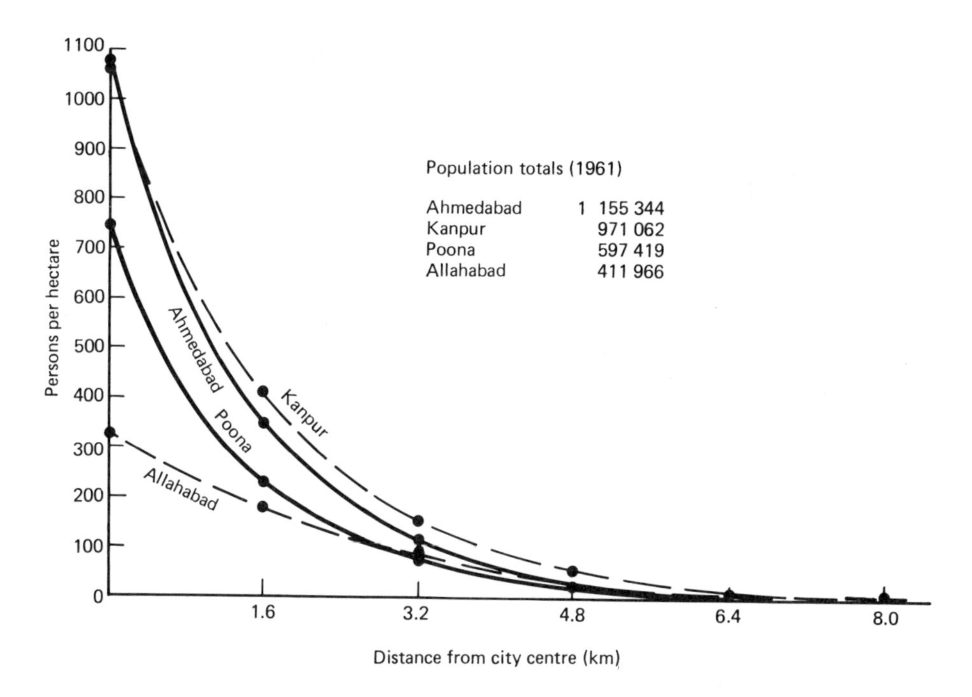

Fig. 8.28 Indian cities: Population density gradients (1961)

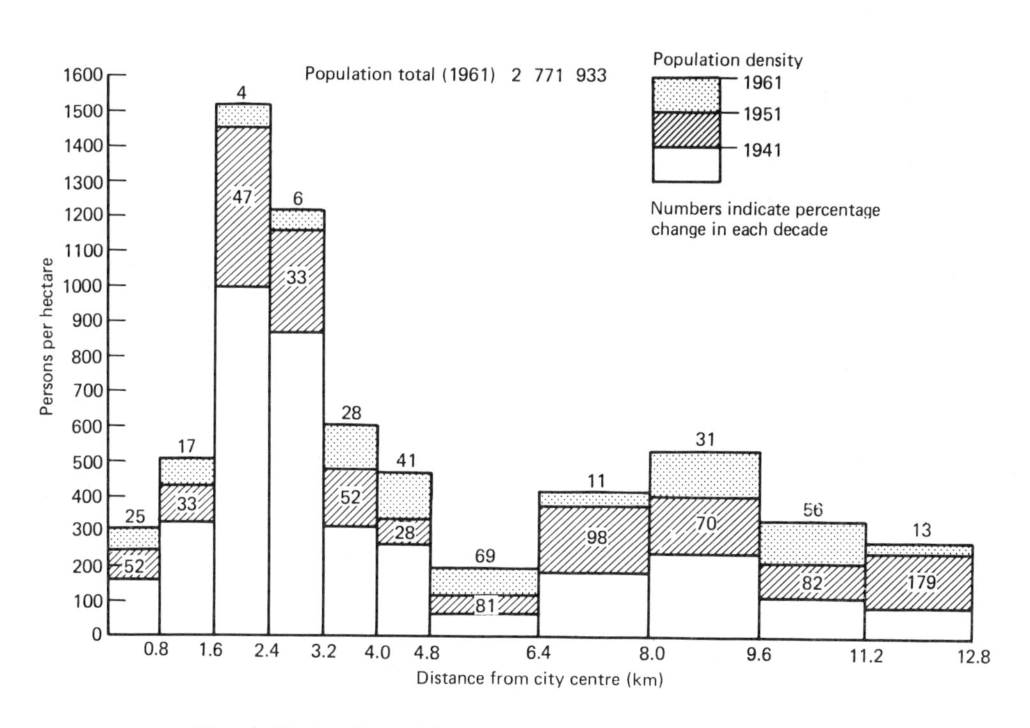

Fig. 8.29 Bombay: Changes in population density 1941–1961

Index